Middleton Goldsmith

A Report on Hospital Gangrene, Erysipelas and Pyaemia

As Observed in the Departments of the Ohio and the Cumberland - with

Cases Appended

Middleton Goldsmith

A Report on Hospital Gangrene, Erysipelas and Pyaemia
As Observed in the Departments of the Ohio and the Cumberland - with Cases Appended

ISBN/EAN: 9783337161323

Printed in Europe, USA, Canada, Australia, Japan

Cover: Foto ©berggeist007 / pixelio.de

More available books at **www.hansebooks.com**

A REPORT

ON

HOSPITAL GANGRENE,

ERYSIPELAS AND PYÆMIA,

AS OBSERVED IN THE

DEPARTMENTS OF THE OHIO AND THE CUMBERLAND,

WITH CASES APPENDED.

BY M. GOLDSMITH,
Surgeon U. S. V.

PUBLISHED BY PERMISSION OF THE SURGEON GENERAL U. S. A.

LOUISVILLE:
BRADLEY & GILBERT, CORNER OF THIRD AND GREEN STREETS.
1863.

REPORT.

Louisville, Ky., Sept. 1st, 1863.

To Brig. Gen. W. A. Hammond, Surg. Gen. U. S. A., Washington, D. C.:

Sir—
The investigations which have been conducted in this city in relation to hospital gangrene, erysipelas, ichorrhæmia, thrombus, metastatic abcess (pyæmia,) diphtheria, and gangrenous scarlatina, have had for their objects to ascertain:

First, the nature of the causes operating to the production of the diseases; Secondly, the nature of the process set in motion; Thirdly, the prophylactic and curative agents.

The affections just named are, in some of their aspects, so intimately united on the points touched in these investigations that it is impossible to disunite them for the purpose of description.

The opportunities presented in the Military Hospitals of this city for the study of some of these diseases have been great, especially so in regard to hospital gangrene. For the study of diphtheria and scarlatina, except in connection with wounds, the occasions have not been numerous in these hospitals, the cases having been isolated and few. Indeed more cases have occurred in private than in public practice. Enough, however, have been observed to justify the conclusions expressed in the following pages. The report of the cases of scarlatina and diphtheria will form the subject of a separate paper.

Hospital gangrene, erysipelas, gangrenous diphtheria, and scarlatina, viewed as local diseases, present, on careful study, many points of resemblance; nor are they unlike in the constitutional states wedded to them. These resemblances are marked.

In diphtheria we notice the transformation of the exudate into a diffluent pulp, the erosion of the subjacent tissues, and the exuding of erosive sero-purulent fluid, excoriating the skin and mucous membranes. In scarlatina, we observe the production of sloughing, or ulcerous surfaces, exuding also corrosive sero-purulent discharges. Both diseases evolve a disgustingly putrid odor. Hospital gangrene always presents pulpy, diffluent sloughs; exudes thin sero-purulent discharges, corroding the skin where they overflow the confines of the sore; and emits a pungent putrid odor. Erysipelas, when it runs on to the destruction of the cellular tissues, is attended with like phenomena.

Erysipelas and hospital gangrene would seem to be cognate; and although, in some of their clinical aspects, they do not show a great similarity, yet a careful study of the following columns, setting forth the points of similarity and dissimilarity, will convince the reader that they are more akin than appears at the first glance:

ERYSIPELAS.	HOSPITAL GANGRENE.
1st. Attacks skin and cellular planes (succulent tissues.)	1st. Attacks skin and cellular planes (succulent tissues.)
2d. Does not readily involve tendons, fasciæ, &c., (dry tissues.)	2d. Does not readily involve tendons, fasciæ, &c., (dry tissues.)
3d. Spreads along the skin and cellular planes.	3d. Spreads along the skin and cellular planes, most readily and rapidly.
4th. In the cellular variety presents necrosis of cellular substance, in diffluent sloughs.	4th. In all tissues which it attacks presents pulpy, diffluent sloughs, (except in bones?)
5th. When attended with necrosis of the cellular tissue, exudes corrosive fluids. The yellowish fluid produced in the phlegmonous variety contains no pus or other cells; the	5th. Exudes corrosive sero-purulent discharges, excoriating the skin. The yellowish fluid, which can be pressed out of gangrenous sloughs and sores, contains no pus or other

constituents seem to be fine granular matter, the debris of connective tissue, and a few fibres of inelastic fibrous tissue.

6th. In the cellular variety, emits putrid odors.

7th. Is contagious and infectious.

8th. Is frequently complicated with thrombus.

9th. Leads occasionally to ichorrhæmia.

10th. Is producible by inoculation with putrefying animal matter, as well as some of the the gaseous products of putrefaction.

11th. Arises in over-crowded and ill-ventilated wards.

12th. The empirical remedies addressed to the constitutional state are: Tinct. mur. ferri, quinine, alcohol, etc., (antiseptics.)

13th. Local remedies are: iodine, creosote, tinct. mur. ferri., sulph. ferri., etc., (antiseptics.)

cells; the constituents seem to be fine granular matter, the debris of connective tissue, and a few fibres of inelastic fibrous tissue.

6th. Emits putrid odors.

7th. Is contagious and infectious.

8th. Is frequently complicated with thrombus.

9th. Leads almost constantly to ichorrhæmia.

10th. Is producible by inoculation with putrifying animal matter, as well as by some of the gaseous products of putrefaction.

11th. Arises in over-crowded and ill-ventilated wards.

12th. Empirical remedies addressed to the constitutional state are: Tinct. mur. ferri., quinine, alcohol, etc., (antiseptics.)

13th. Local remedies are: nitric acid, creosote, chlorides, charcoal, &c., (antiseptics.)

In the present state of our knowledge, we regard the local manifestations of erysipelas, of scarlatina, and of diphtheria, as being preceded by and as depending upon certain blood states. In this aspect some resemblance is noticeable: all three of these diseases seem to have their characteristic expression on the tegumentary surfaces.

Erysipelas is no unfrequent complication of both scarlatina

and diphtheria. They are all adynamic diseases, and present to chemical examination analogous changes in the blood.

Of the causes which operate in producing erysipelas, this much is known: Erysipelas is often the product of dissecting wounds—of wounds received in skinning diseased cattle, or in skinning the putrefying carcasses of those killed by accident. It is often seen to result, in the form of puerperal peritonitis, from the infection upon the hands of the midwife, as in the historical German cases. It is often associated with injuries and diseases of the bones, especially with caries, a disease remarkable for the persistent fetor of the discharges. Indeed all that is known with regard to the artificial production of erysipelas may be summed up in the agency of putrefying animal matter.

Erysipelas, too, seems to be engendered in over-crowded and ill-ventilated apartments, reeking with foul emanations of the human body; in rooms receiving exhalations from drains and cess pools, especially from the former. It is producible by miasm emanating from the bodies of those having the disease. The bodies of those having erysipelas, in some of the worst epidemics of the disease, have been known to emit a putrid odor.

The bodies of those who die pass quickly into putrefaction; and in the blood of such, there is often found lactic acid, a product of putrefactive decomposition. The whole drift of testimony goes to show that erysipelatous diseases are more or less connected with putrefactive processes; and although it may not be possible, in the present state of our knowledge, to establish the precise relation between the two, enough is known to make it certain that an intimate relation exists.

Of the causes which produce hospital gangrene, this much is known: Hospital gangrene arises spontaneously in wards where the wounded are crowded together—where the wards are filled with the stench of traumatic profluvia, and receive the air of sewers and cellars.*

---

*I beg leave to interpolate here an extract from a report made by me to the

It is produced by inoculation; and there is reason to believe that it is producible by keeping the putrid flesh of healthy animals in contact with a wounded surface. The poison spreads through the medium of the atmosphere, and adheres with great endurance and tenacity to fomites. That the gangrenous process—the spread of the gangrene—is propagated by the ever produced new gangrenous matter, is obvious to the most superficial examination.

As to the mode of operation of the agents producing erysip-

Surgeon General in regard to the indigenous production of hospital gangrene in the hospitals at Nashville: "I find that in Hospital No. 8 there have occurred, according to the best evidence in reach, thirty-eight (38) cases of hospital gangrene, of indigenous origin; meaning, by indigenous origin, those cases not produced by infection, as where a man having been brought into a ward with hospital gangrene imparts the gangrenous process to the wounds of other men.

The facts are briefly these:

1st. All of the cases occurred in Ward No. 1.

2d. All the cases occurred in the row of beds next the windows opening on the alley.

3d. All the cases occurred prior to the 24th of April, or during the time when the external atmosphere was colder than that of occupied houses, closed cellars, or underground drains.

4th. The cellar under the hospital had passing beneath it, and opening into it by several apertures, the common sewer of that part of the city.

5th. The soil pipes from the privies of the several wards traversed this cellar, and emptied without a trap into the common sewer.

6th. This soil pipe is made of tin and leaks badly.

7th. In wet weather the cellar bottom is overflowed by the contents of the soil pipe and sewer.

8th. This cellar has but two openings, one in the front of the building and one in the alley.

9th. The alley is long, narrow, and the hospital buildings and barracks high.

10th. The area of the adjacent building receives the drainings of the garbage of its kitchen, and this area forms a part of the alley.

11th. Ward No. 1 derives its ventilation almost entirely from windows opening on the alley. On the opposite side there is but one opening, a door leading to a hall which has no window. On the end next the street there are but three windows.

12th. The prevailing winds during cold weather sweep the street in the front of the building, leaving the atmosphere in the alley almost undisturbed.

13th. The emanations from the area of the adjoining building, as well as those from the cellars of the Barracks and Hospital are most offensive at all

elas and hospital gangrene, but little is known accurately. The erysipelatous virus, waiving all considerations as to its personality, seems to operate through the atmosphere, and, according to traditional views, gains entrance into the blood through the respiratory organs. This subject needs reinvestigation, for there are some facts which seem to militate against this received view of the matter. These facts are the following:

times, and are most disgustingly perceptible in the evening when the external atmosphere begins to grow cool.

14th. The cases of hospital gangrene did not recur after the weather grew warm—when the outer air was warmer than the air in the cellars or sewer.

15th. The buildings on the opposite side of the street (the alley running through only one square,) prevented any wind from traversing the alley below the second story.

In the early part of the season one patient was brought into the ward with hospital gangrene. In a few hours six other cases were developed in wounded men lying adjacent to him. The disease did not spread, and, with the termination of these cases, disappeared for a season. When it broke out again, it attacked those who had come into the Hospital without any appearance of a gangrenous condition of their wounds at the time of admission.

It will appear, from the facts related, that the miasms evolved by putrifying animal matters in the cellar of the hospital, and perhaps in the area of the barracks, were given off at all seasons; that just during the season when, from the relations of the temperature, the atmosphere of the cellar and this upward current would enter the ward most constantly (*i. e.* when it was necessary to heat the ward with stoves,) the cases of gangrene occurred; that they occurred in just that locality in which the miasm of the cellar, in least dilution, would impinge upon the patients; and that as soon as those relative conditions of the temperature obtained in which the miasm of the cellar would flow downwards (*i. e.*, when the external air was warmer than that of the cellar,) and when, from the extinction of fires, no air was drawn in at the windows, the disease ceased.

The testimony of the Surgeon having the ward and cases in charge is all the more valuable for that he did not anticipate or interpret the facts. The Surgeon noticed without peculiar concern the occurrence of the affection, and, as he expressed himself to me, supposed that by some singular chance "those patients having the lowest vitality were placed in that row of beds." And he marveled greatly that the cases should occur just where the ventilation was best, because the windows on the alley, being the only available inlets for air in the ward, were opened diligently four times a day, and some part of each window was kept open all the while. I think that the records of surgery do not afford a more unique or striking example of one of the methods of the production of hospital gangrene."

It is found that those recovering from erysipelas of the head and face are, after desquamation of the face, peculiarly liable to new attacks on exposure to the miasm in infected wards, and that this liability to the disease is prevented by coating the face with a tincture of iodine, (which process detaches an epithelial coating,) or by constant embrocations with glycerine, or with castor oil, with collodion, &c. Erysipelas, attacking the face, respects, in its first invasion, the parts which by accident are covered by such coatings as are made with adhesive plaster. It hardly ever *begins* in parts covered by hair and sebaceous matter, as the scalp.

If the poison of erysipelas acts from the blood, it is difficult to understand why it so constantly selects the face as the beginning point, and yet commencing in the face extends so rapidly to the scalp, and neck, and thorax.

Any part of the skin, provided only that it is abraded, may be invaded by erysipelas. That of the trunk or extremities is attacked, provided there is an ulcer or wound. If a man has caries, or necrosis, or hospital gangrene, or any disease commonly associated with erysipelas, the latter attacks the skin at the site of the original disease. Now a morbific agent in the blood is said to attack those tissues or parts of the body which have a supposed affinity for the poison, a susceptibility or sensitiveness to it: as, for instance, ergot impresses the uterus; phosphorus, the jaw; nitrate of silver, the skin and kidneys; but it is not known that any producible condition of these organs renders them any more sensitive to the matters circulating in the blood. The precipitation of the silver on the particular tissues seems to depend upon original endowments of the parts. How is it with the poison of erysipelas? The theory runs that it enters the blood, and, by the operation of a sort of elective affinity, it attacks the skin of the face, and respects the skin of all other parts in relation to points of original invasion. True, beginning in the face, it may spread to other parts, but it (idiopathic erysipelas) rarely begins at any other point except as mentioned hereafter. The non-parturient woman does not have erysipelatous metritis or peritonitis, however much she is exposed to the inhalation of the miasm; if she has erysipelas at all, she gets it on the face, like other people. Men and non-parturient women breathing the miasm get the disease on the face, not on the trunk or extremities. But let a woman give birth to a child, and, during the parturient state, let her be exposed to the miasm, and straightway she gets puerperal peritonitis. During the non-parturient state a hundred women may be exposed to the miasm, and not five per cent. will have erysipelas of the face; and yet, if one hundred parturient women are exposed to the miasm, how many would escape puerperal peritonitis? Would twenty per cent.? Now why is the parturient woman more influenced by the miasm? Certainly not by reason of any blood state; for all the blood changes peculiar to her condition have their climax on her delivery, and exist in large development anterior to her delivery. The miasm seems potent not prior to but after delivery, when the puerperal blood states are in the process of involution. What is the pathological condition of the parturient woman precisely coeval with the period in which she is so exceedingly liable to the invasion of erysipelas? Just this: her uterus is stripped of its lining epithelium; the internal surface of the uterus is like a piece of skin stripped of its epidermis. After this epithelial covering is reproduced, the parturient woman is no longer any more liable to the invasion of erysipelas than she was before she became pregnant.

Again: expose a hundred men to the miasm of erysipelas; a certain number will have erysipelas of the face, the disease always commencing in the eye-lid, in the alæ of the nose, or behind the ear. The disease commences in no other part of the body; it does not attack the trunk or extremities. But if wounds or abrasions exist, the erysipelas no longer selects the face, but attacks, without discrimination of region, the parts wounded or abraded—provided only that the parts thus wounded or abraded are uncovered.*

* I here interpolate another passage from my report to the Surgeon General on the indigenous origin of hospital gangrene and erysipelas:

This is, at least, the general rule. True, that wounded men sometimes get the erysipelas in the face, whilst the wounded parts are respected; and, as in some instances observed here, after the erysipelas has run its course in the face, the wounded parts have become involved. Still the general rule is, that, in idiopathic erysipelas, the disease begins in the face, except in children, and the exception is a pregnant one. Children have idiopathic erysipelas in the extremities almost as often as in the face, but this liability to erysipelas in the legs of children is confined to the period of long clothes; the disease no longer appears idiopathically when the children are weaned, and advanced to short clothes.

Again: those who have had erysipelas of the face, and who have just recovered, are, as is well known, exceedingly liable to new attacks, relapses, especially in hospitals. Now this liability disappears—the men are protected by coating the face with tinct. of iodine, or by frequent application of glycerine, or simple cerate, or resin ointment.

"The other occurrences which I wish to relate were observed at Hospital No. 3. The building is five stories high, and consists of two stores; it stands upon the corner of the streets. The sides of the building fronting on the streets have windows in them. The other sides have none, except one in each ward at the end, where it abuts for two stories upon an alley or area about six feet wide; a hoistway is cut from the lower to the upper ward in that half of the building under consideration. During the continuance of erysipelas at this hospital, all the cases occurred in the upper ward of this side, and the disease continued unabated until the glass in the skylight overhanging the hoistway was broken out, so as to allow the foul air ascending from the lower wards to pass through the roof; cotemporaneously, three windows were opened in the side of the ward, and two in the end.

The inmates of this ward were mostly and during a part of the time exclusively wounded men. The singular fact noticed in regard to the cases was that the erysipelas in no case attacked the wounded parts. It always attacked the face. Such an occurrence demanded investigation. I enquired of the Surgeon if he pursued any peculiar treatment. The Surgeon in charge, who was standing by at the the time, remarked, with a smile, that the Assistant Surgeon in charge of the erysipelas ward had a peculiar and uniform way of dressing a wound, a way which he never departed from; it was this: *He always, after having cleansed the wound, covered it over completely with ung. resinæ, so placed as to exclude the air from the granulations completely.* After this dressing, warm water was applied over the wounded parts.

Further: all the known blood poisons—all foreign substances tending to precipitation in certain parts, elimination by certain organs—seek, if the habitual destinies of the substances are interfered with, vicarious locations or secretions. Thus uric acid, in gout, tends to certain points, *e. g.*, the joints; now if the deposition of uric acid or its compounds at the joints is interfered with, straightway the uric acid seeks another nidus. If the agent producing rheumatism is driven away from its resting place, it straightway seeks another. If bone-earths, redissolving in certain bone diseases, fail of elimination by the kidneys, they are deposited in the lungs or in the coats of the stomach. No local treatment of gout or of rheumatism aborts the disease. The utmost hope of topical treatment is palliation of pain, or of local pathological processes. The agents here are known and recognizable blood poisons. The local affections continue because of the continuous application of the morbific agent. With the expurgation from the blood of the agent, with the drying up of the fountains from which those poisons flow, the local diseases cease—and all remedial means look in this direction.

The records hereto appended show that erysipelas may be interrupted at any point in its progress or development, and that any such interruption does not produce metastasis; nay, more, it shows that the constitutional state not only is not aggravated thereby, but that, in the obsolescence of the local disease, the constitutional symptoms vanish.

Further, all that we know of the order of events in inoculable diseases—as vaccinia and hydrophobia—and their abortion, corresponds with what we claim for erysipelas.

In vaccination we have a period of incubation of about four days; at the end of this time there is some elevation of the skin, but no redness—this corresponds to the period of infiltration. After the fourth or fifth day, inflammation sets up, and coincidentally the symptomatic fever. If, at this point, the vaccinia is aborted by any accident, the constitutional state falls with it; nor is the protective agency of the vaccination developed.

May it not be possible that the poison of erysipelas acts in the same way? that there belongs to it a period of local incubation—zymosis—corresponding with the period of infiltration in vaccinia? that then the inflammatory stage is developed; and that constitutional infection may be produced, at any period intervening between the first contact of the infectious matter and the development of the local process, as well as after the inflammatory symptoms are developed?

Now if the ancient dogma is founded in truth, and if the miasm of erysipelas gets into the blood, and by virtue of hypothetical affinities precipitates itself upon the skin of the face, as nitrate of silver does upon the whole skin, and as ergot does upon the uterus, we have to assume, in order to make the dogma coherent, that wounds endow the skin of the extremities with a new order of affinities; that the puerperal state of the uterus endows it and the peritoneum with new sensibilities; that the skin on the legs of infants has affinities that disappear with age; that the coating of the face in the adult with grease destroys its affinities—and all this in the face of the fact, that wounds, the puerperal state, age, and protection, are not known to have any influence upon these affinities or local sensibilities in connection with measles, or scarlatina, or small-pox, or with other diseases affecting the skin!

Do not these facts quadrate better with the idea that the contagion of erysipelas acts rather by contact with the skin? That the contagion, floating in the atmosphere, attacks the face because the face is the part uncovered? That the face is attacked, in preference to the hands, because of its thinner epithelium and more superficial vessels? That the disease attacks the eye-lid, the alæ of the nose, and the skin behind the ear, because just precisely here the epithelium is thinnest? That it avoids the hands on which the epithelium is thick? (Whoever saw erysipelas attacking the unwounded palm of the hand, or unwounded sole of the foot?) Would it not be reasonable to suppose that desquamation of the skin or its division by wound or ulceration—the exfoliation of the maternal surface of the uterus—opened the parts to the action of the contagion just as the same conditions would open the parts to the action of medicine? That the constant exposure of the child's legs to the contact of air, (or that perhaps the fact that in the child's body there are not yet developed those differences in the thickness of the epithelium which obtain later in life), allows the contagion to act with equal advantage upon different parts of the body? That the protection of the tender and thinly covered skin of those just recovering from erysipelas of the face may prevent the action of the contagion, and thus give immunity to relapses?

In all this I do not claim the abnegation of the traditional dogma, but I think the considerations here urged, and the records hereto appended, show sufficient ground for demanding of thinking men the re-investigation of the accepted theory that they show, at least, that the increasing gravamen of some of the diseases, supposed to originate in blood infections, is due rather to the retroactive influence of the local parts of those diseases than to an increasing quantity or intensity, or to a growing zymosis, of the original blood poisons.

Pyæmia is divided into two conditions; that is, the conditions formerly expressed by this term are now known to be: first, a state denominated ichorous infection—ichorræmia—a condition pronounced commonly, when exquisitely developed, by frequent rigors, feeble pulse, loss of appetite, colliquative local sweats, a fermentative odor of the breath, pallor of the lips and face, and great muscular prostration; second, metastatic abscesses and inflammations, produced by the softening and detachment of thrombi previously formed.

Both of these states may or may not be complicated with leucocytosis, primarily or secondarily.

The first of these conditions, ichorræmia, is the constitutional state developed with the greatest rapidity, most rapidly leading to a fatal result, and connected more frequently with gangrene and with injuries, especially those involving bones.

Thrombosis, the commencing point of the second class, is frequently, not always, associated with ichorræmia. Both of these processes, the ichorræmic and the thrombic (pyæmic), are connected occasionally with the more serious local effects of erysipelas, as abscesses and sloughing of the cellular tissue; on the other hand, erysipelas and hospital gangrene are not necessarily followed by ichorræmia, although but few cases of hospital gangrene present themselves in which more or less of ichorræmia does not exist, at least during the advance of the gangrenous processes.

This ichorous infection seems to depend upon the presence of putrid matter in wounds, in suppurating cavities and gangrenous surfaces or excavations. The record of cases will show that a constant correspondence exists between fetid discharges and ichorous infection. In all the cases carefully noted in this context, it has been observed that the discharges were fetid prior to the development of the symptoms of ichorous infections. The fetor may have been noticed for variable periods, but was always noticeable prior to the constitutional symptoms. Ichorous infection does not always follow upon fetor of the discharges, but ichorous infection never occurs where the discharges are not fetid; not only so, but in all of the cases recorded it will be seen that, as soon as the fetor was corrected, the symptoms of the ichorous infection began to disappear, and that, too, with a rapidity which was surprising to all the observers.

Again: it will be seen that relapses took place, and that relapses were coincident with a return of the fetor.

The constant co-existence of the putrescence of the products of ulcerous surfaces, and ichorous infection—the constant precedence of the former in point of time, the constant disappearance of the latter upon the disappearance of the former, point out an intimate connection between the two events; indeed, establishes the dependence of the latter upon the former. What the peculiar substance is, which, getting into the blood, gives rise directly or indirectly to the ichorræmia, we do not as yet know. Of it, we know this: first, that it is developed during the putrefactive decomposition of animal matter; second, that it is not pus in its cells, intracellular fluid, or intercellular fluid, (I refer to normal and so called reproductive pus); third, that it is a fluid, or dissolved in a fluid. That it is developed during the putrefactive decomposition is made probable by the following facts: ichorous infection, as stated above, occurs only in cases where this process is going on; the effects of the agent disappear upon the arrest of putrefaction in the wound; the influence of the agent is destroyed by any substance which arrests putrefaction. Further: somehow empirical medicine has fallen into the habit (and with good reason, doubtless) of employing certain remedies. These agents have, traditionally and in truth, value in the treatment of the several affections. All that we know of this treatment is summed up in these empirical remedies. Let us examine them.

First, as to erysipelas: the topical remedies known to be most efficient by virtue of any active property of their own, are, tinct. iodine, tinct. mur. ferri, creosote, and sulph. ferri. These substances are diverse in their active properties; in what does creosote resemble iodine or iron? They agree only in this, that they are antiseptics, that they break up the putrescent actions. Secondly, as to hospital gangrene: The local remedies growing into repute, after long and disheartening experience, are nitric acid, nitrate of silver, acid nitrate of mercury, arsenical solution, chlor. of soda, and the actual cautery. These are all antiseptics. They agree in this, and in this alone—that they break up the putrescent series directly or indirectly. Thirdly, as to pyæmia: All the agencies found to be useful in preventing the development of pyæmia are free openings in depending positions, drainage tubes, and cleanliness. Some use has been made of injections of chlorinated water, and of solutions of hyposulphite of soda and of iodine. These look in the same direction: on the one hand, cleanliness, depending openings, and drainage tubes, look to the easy and rapid removal of the discharges of wounds; and on the other, iodine, chlorinated washes, and the sulphites, to the arrest of putrefaction in animal matter. It would seem a little singular that, in the slow and gradual growth of empirical medicine, remedy after remedy, selected by purely experimental processes, and not under the domination of any theory as to the diseases in question, should be found side by side in common use, and that those remedies should be found, however diverse they are in their other qualities, to agree with perfect uniformity in one common quality, unless that one quality had some relation to the special nature of the diseases or the causes of their production.

A series of investigations is now on foot seeking to discover the essential agent, if one exists, which sets up these diseases. It may be that there is one product of the putrefactive decomposition appropriate to the production of each of these diseases. It may be that the putrilage, or some of its associate products, exercises a catalytic influence in the production of the local processes, and that the putrid fluids, passing into the circulation, exercise a direct poisonous agency, or set up in the blood analogous processes. It is probable, for the results thus far reached point in this direction, that peculiar alkaloids are produced in putrefactive decomposition, and that one or more of these are the active agents. Whatever the agents are, they exist to some extent in all kinds of putrescent animal matter.

While it may be premature, at the present stage of the investigations which I am engaged in making under your orders, to attempt to determine the agent specially concerned in the production of pyæmia, this much, I think, can be stated as probably true:

1st. That it is not the final product of putrefactive fermentation. The final products of fermentation or putrefaction differ from the initial and mediate products. As in the fermentation of starch we have alcohol and vinegar, so in the putrefaction of animal tissues we have, according to the stage of the putrescent motion, differing products. So, again, according to the conditions of moisture, heat, and access of air, we have putrefaction going on with different phenomena and products. Who has failed to perceive the differing results of that putrefaction which takes place in the ordinary process of maceration, in the cadaver enclosed in a metallic coffin, and in a carcass exposed to the sun?

2d. The agent will probably be found to be generated most freely in the initial stages of putrefaction. This view is borne out by the results of injections into the veins of animals. Solutions of putrid pus, which do not evolve ammoniacal products, freely produce commonly a group of symptoms more allied to ichorræmia; while solutions of highly putrid substances commonly produce death with great rapidity, engendering gangrenous affections, or in less degree producing diarrhea, vomiting, or a condition allied to the ammoniæmia of Jacksh. The discharges and gangrenous pulps, so far noticed as connected with ichorræmia, had an acid reaction, always disappearing on keeping the specimen until ammonia was developed freely. At present, however, nothing definite or certain in this relation, other than the meagre facts above stated, can be adduced. This branch of the subject is under investigation by two chemists of ability, one here and one at Nashville.

The difference in the resulting symptoms seems to have some relation to the physical characters of the putrilage: thus, the more fluid the products, the more readily the constitutional symptoms seem to be produced—the more coherent, the drier the products, the more do the effects appear in local processes; at least such are the impressions I have received from the observations so far made.

In regard to the directly appreciable effects of the gangrenous putrilage of hospital gangrene, the following observations have been made: The first effect is the coagulation of blood in the blood vessels on or near the surface to which it is applied. This coagulating power manifests itself sometimes in the blood of large and thin-walled veins, seldom in the large arterial branches. A few instances have occurred in which the saphena, or its branches, have been involved in the progress of the gangrenous chasms. In such instances it was noticed that the vein, if destroyed at all, was destroyed very slowly, but that the first change was the coagulation of the blood in the vein. In a few cases the vein, though exposed and bared of its sheath, was not destroyed. In such cases the vein was almost uniformly felt to be hard, whip-cord-like, the induration beginning at the distal margin of the excavation and extending a little beyond the cardiac margin. The hemorrhages which have occurred have all been arterial. This coagulating power of the putrilage seems most marked in the thin-walled, deep-seated veins—least marked in the thick-walled, muscular, superficial veins. In the smaller veins the coagulation is complete. In the larger veins the coagulation takes on more the character of growing thrombus. Now it is a noticeable and pregnant fact, that the thrombi which were found to be formed in the larger veins seemed to have their commencement at that point where the gangrenous cavities came nearest in contact with the vein, and the thrombus seemed to start a little heartwards from that point, extending thence toward the heart. The observations on this point have not been numerous enough to warrant the promulgation of this statement in the form of a universal truth, but they tend in that direction.

The fluid parts of the putrilage seem to flow into the cellular planes, inducing in them a sort of gangrenous liquefaction. The tendency of hospital gangrene to spread through cellular planes seems to be explained by this fact. In some of the cases, where the disease had manifested itself on a large surface, spreading with frightful rapidity—cases in which, on the application of our remedies, the disease was promptly arrested—the muscles could be seen as nicely dissected as if prepared for class demonstration by the anatomist's knife.

Occasionally a condition of things is developed in the subcutaneous cellular planes, throughout great districts, in which the cellular substance is seen to grow yellowish, lose its coherence, and, in the more advanced stages, break down into a putrid, yellowish fluid. This fluid, under the microscope, is seen to consist of fine, granular matter, is totally devoid of cells, and presents threads of inelastic fibrous tissue. This diffluence of the cellular tissue eventually becomes completed to such an extent that the fluid products flow from part to part, leaving the superior portion with the skin (of an arm, for example) appearing like a flaccid bag. The skin becomes discolored, but is not thickened and indurated as in facial erysipelas. In some cases the skin is pale, sometimes livid; at certain points, and generally to a large extent, the skin becomes necrotic. The disease spreads in all directions, but chiefly downward. In those who live long, small flocculent sloughs pass away, but the bulk of the discharges is the thin, yellowish, stinking fluid above described.

The predominating symptoms are chills, small and frequent pulse, dry hot skin, dark yellowish tint, red and dry tongue, distressing hacking cough, frequent and superficial respiration, singultus, hebetude, sopor, and delirium. This answers to the so-called phlegmonous erysipelas, and differs from the limited abscesses so common in the eyelids in facial erysipelas. Besides the influence exerted by the putrilage upon the blood contained in the vessels, it starts, in the tissues with which it comes in contact, the putrefactive process. No transitional state of the tissues touched has as yet been observed. The gangrenous and living parts are shaded—blended into each other—so blended that it is not possible with the naked eye to see where the dead tissue ends and where the living begins. The tissues soon deliquesce, but this deliquescence is not progressive; the older parts of the sloughs do not appear to be materially softer than the newly forming sloughs. Some of the elements of the tissues seem almost always to afford a dirty, yellowish fluid, a fluid which can be readily expressed from the coherent parts of the slough.

This fluid contains no corpuscular bodies, is not separable into a supernatant serum and sediment by standing; it has no quality of pus, except its fluidity and color. Pus corpuscles have never been seen in the products of a gangrenous sore, unless at some point granulation was going on. From all that can be seen, the parts die—are destroyed in pretty much the same manner as they are destroyed by continuous rubbings of a tumor with caustic potash.

It will be seen that, in my opinion, at least, these three affections, hospital gangrene, pyæmia, and erysipelas, are in some way connected with miasms, or with poisonous substances, by some process developed in animal matter in the course of a series of chemical actions expressed in the generic term, putrefaction. In consonance with these views, and, indeed, under the domination of this idea, the search was instituted, having for its object the discovery of an agent possessed of the power of arresting putrefaction and of destroying the products of putrefaction in whatever form those products might present themselves, solid, fluid, or gaseous. Just at this point I had to deal with the difficulty which presented itself in the poverty of our knowledge of these putrefactive processes, their products and effects upon the living organism.

In order to be able to state the exact extent of our knowledge of this matter in its chemical aspect, I here insert a note from Acting Assistant Surgeon Jenkins, an expert chemist, charged with the investigation of the diseases in question, so far as appertains to the application of practical chemistry.

"LOUISVILLE, KY., April 2d, 1863.
"DR. M. GOLDSMITH:

"*Dear Sir*—In reply to your inquiry of a few days ago, asking me to state what is known at the present time as to the chemical nature of the causes producing or the products resulting from infectious diseases, gangrenous ulcers, etc., I think I can safely say that little or nothing is known as to the real and substantial causes of such morbid processes; and in reference to the nature of the products emanating from ulcers, etc., about all we know is this: the effluvia from foul and putrescent ulcers are composed chiefly of sulphuretted and phosphuretted hydrogen, ammoniacal and sulphuretted ammoniacal gases, with a little carburetted hydrogen. All of these bodies have been proved to be incompetent to induce diseases similar to those which gave rise to them.

"Among the many substances which have been employed with a view to the arrest of the disease, or the checking or prevention of putrefaction, or the destruction of the noxious vapors and gases, may be enumerated both physical and chemical agents, such as charcoal to absorb, balsamic fumigations to disguise, and antiseptics to change the chemical composition of the volatile products of putrefaction—for example, chlorine, some oxy-acids (hyponitric acid, sulphurous acid).

Some salts, (nit. of lead, sulphate of iron,) again act by a thoroughly destructive effect upon the substance of the affected tissues, and destroy the integrity of the organized structure, *e.g.*, corrosive acids, caustic alkalies. Others, again, act in a milder manner, and appear only to coagulate the albuminous fluid in and about the diseased parts, *e. g.*, corrosive sublimate, nitrate of silver, etc. As an example of what we know in reference to the action of antiseptics, such as chlorine, nitrous acid, sulphurous acid, etc., I will state that they decompose the organic body giving rise to offensive and noxious emanations, and act as disinfectants and antiseptics,

"1st. By abstracting water.

"2d. By forming with organic matters compounds less susceptible of decay.

"3d. By deodorizing the body.

"4th. By destroying cryptogamic plants and infusorial animalcules.

"Little, also, is known upon this subject, of a definite nature, as most of the experiments have been made with a view not to strike at the cause, but to destroy the noxious character of the products of putrid disease.

"Very respectfully yours, &c.,
(Signed,) "THOS. E. JENKINS."

In looking over the lists of the agents already in use in gangrenous affections it was found that the corrosive acids and the actual cautery, the most potent to arrest the gangrene, involved the large destruction of living tissues, and therefore were not applicable to the arrest of the process, except upon open surfaces. Their operation is limited to the mere charring of the tissues. Neither could be brought safely into gangrenous caverns, and both, in their clinical application, are attended with so many disadvantages that they have not completely answered the purposes in view. As to the other agents in vogue, although operating in the right direction, they are ineffectual, confessedly so—a fact most eloquently attested in the death-rate which attends upon the disease.

My attention was at once turned to the halogens, and to substances acting like the halogens. The halogens, as is well known, are fluorine, chlorine, iodine, and bromine. The action of these substances is alike in kind, but different in degree. They differ in their physical properties. Fluorine cannot be isolated on account of its ravenous activity. Chlorine cannot be had pure except as a gas, or as a fluid, unless under pressure, and all the solutions of chlorine are feeble in their action. Iodine cannot be had at ordinary temperatures in a gaseous state, and cannot be employed except in a solid state, or, if in a fluid state, too largely diluted to possess the necessary degree of activity.

Bromine was found to be a fluid vaporable at ordinary temperatures, and, unlike chlorine, respirable without injury or inconvenience.

Dr. Brainard, of Chicago, had, as I am informed, already found in his researches upon the bite of the rattlesnake, that iodine mixed with the virus destroyed its activity, and had shown that like effects were produced upon other animal poisons. From these facts, bromine seemed to hold out the most promise for the purposes in view. Reasoning from the active antagonism known to exist between the halogens and animal poisons in substance, the inference seemed reasonable that a gaseous halogen would be antagonistic to vaporous animal poisons. Indeed, this quality had already been proven to exist in chlorine. The objection to chlorine for the purposes of disinfection, in wards occupied by the sick or well, is that very small quantities of it render the atmosphere irrespirable; and, as the amount requisite was found to be larger than was compatible with respiration, no trials were made of it.

The first occasions for the experimental use of the agent selected were in connection with erysipelas, its prophylaxis and treatment. Finding, as will be seen by the records appended, that the indications were fulfilled beyond our most sanguine expectations, the bromine was supplied in large quantities, and held ready for use in hospital gangrene should it appear in our hospitals. The records in regard to its effects in this disease are herewith appended.

Investigations are now on foot in regard to pyæmia, and the use of the bromine and its analogues in prevention. The cases apt to be observed and treated with this agent have not been sufficiently numerous to develope well-grounded results. They promise still more important results than have been reached in the treatment of erysipelas and hospital gangrene—more important only, however, from the fact, that pyæmia is more frequent in its occurrence, and destroys more lives than both the other diseases.

The cases of hospital gangrene, which have been treated in the hospitals at this place, have presented some constancy in most of their characteristics; and in order that the true nature of the disease may be apparent to the Surgeon General, I will state the appearances commonly presented: 1st. The gangrenous affection presented a tolerably constant tendency to the assumption of a somewhat circular form. This tendency, however, was frequently interrupted by the varying effects of the special remedy used. Thus, when the skin was not undermined to any great extent, the disease was commonly arrested immediately, and the ulcer left presented the usual circular form; but when the skin was much undermined, or undermined to unequal extent, at different points in the circumference, the disease was not arrested as promptly at one point as at another, and thus, eventually, the circular form was lost. Then, again, the gangrenous erosion had sometimes an irregular elongate form, coinciding with the original shape of the wound. Sometimes the gangrene attacked the walls of a wound passing through a limb, and presented itself as the sloughing core of a ball wound, the pulps protruding from the apertures of both entrance and exit.

2d. The spread of the gangrene seemed pretty generally to be influenced by the succulency of the tissues. Thus, when it commenced on the surface in a superficial wound or ulcer, it generally spread most rapidly in the skin and cellular planes. The disease continuing, the muscular substance would next be attacked. Dense fasciæ, as the fascia lata, and tendons, resisted the influence much longer; on the whole, the bones suffered more than tendons. Another reason why the disease spread laterally rather than deeply, and of consequence more in cellular planes, and less rapidly towards the deeper tissues, is, that the sloughs could be more readily detached from the exposed parts of the gangrenous surface so as to allow the effectual application of the remedy, while it was difficult to detach the sloughs underlying the skin, and, of course, difficult to mix the remedy with the sloughs—the *sine qua non* of its curative agency. These causes have operated so decidedly in modifying the form and appearances of the granulating sores left on the subsidence of the gangrene, that many of the numerous surgeons visiting this place, to see for themselves the effects of the treatment instituted, have found it difficult to realize that they were looking upon the ravages of hospital gangrene, unless it so happened that I could show them some cases in which the disease was in progress.

3d. The sloughs were variant in their consistence, and this variance ran from tolerably firm escars to diffluent pulps. The consistence of the sloughs coincided rather with the consistence of the tissues sloughing. The sloughs of skin were soft, swollen, tolerably coherent masses. The sloughs of cellular tissue were soft, flocculent, yielding more abundantly a dirty yellowish fluid. The sloughs of muscles were firmer, less pulpy, more coherent. In some of the cases, and especially in those in which the process was slower, skin, cellular substance, and muscles, seemed to melt away into mere diffluent matter, the product of the destruction of each of the several tissues in these cases being nearly alike.

4th. The sloughs were commonly of a dirty greyish hue, those of the skin being, in most instances, somewhat darker than those of the cellular substance. The variations of color appeared to be influenced more by the quantities of altered blood in the tissues than by any other condition.

5th. In all of the cases there was present a most pungent and intolerable fetor. In some instances the pungency of the gaseous effluvia was so great as to produce a persistent smarting in the eyes and the nares of the persons engaged in dressing the sores. The odor would often fill the whole ward. This fetor, in greater or less intensity, was the almost constant attendant upon the gangrenous process, appearing when it began, continuing as it continued, and ending when it ended. So constant was this coincidence, that those who treated the cases came to regard the disappearance of the fetor as the reliable evidence of the arrest of the disease; the presence of it, as the signal of the commencement of the process. This odor was peculiar; it was not the sickish odor which is often perceived in suppurating or ulcerating wounds, nor the odor of common gangrene, or of common putrefaction of dead animal matter, but an odor peculiar, recognizable with the nose, but not admitting of description.

6th. An examination of the sloughs under a microscope shows that the fluid products consist chiefly of a granular debris, and of fibres of inelastic fibrous tissue; for the rest, morphological forms of the tissues are gradually lost.

7th. In the cases under observation, the only local condition rendering the parts liable to the invasion of the gangrene was solution of continuity. It will be seen that the disease was developed in sores, small and nearly healed, as well as in those which were extensive and recent; that in one case, especially, it was developed at the site of a purpuric extravasation, and that in another it invaded a point almost completely cicatrized. It invaded wounds recent, wounds granulating, and wounds ulcerating. In some few of the cases the disease could be traced to no contagion; in others it was distinctly traceable to the presence of the disease in other patients. That the disease, when developed, was contagious is shown by the occurrence of several cases in the beds next adjoining those already affected.

8th. The constitutional state existing at the time of the invasion, and prior to the invasion, did not seem to have much influence upon the liability to the disease; for the latter seemed to invade the strong and the feeble, the young and the old, the sick and the well, with equal facility. Nor, on the whole, was it noticed to advance any more rapidly in those of depraved health than in those of sound health. The condition of the health seemed to have much to do with the rapidity and the vigor with which granulation was set up after the gangrene was arrested, but to have little to do with the liability to or the progress of the disease.

9th. The constitutional disturbance wedded to the disease was also variant in degree. The quality of the disturbance was pretty uniform. The more noticeable symptoms were: frequency of the pulse, running, in some cases which I observed personally, to 150 per minute—the pulse was also feeble and small; the appetite in every case was diminished; in some, lost; as a general thing, the patients expressed a disgust for meat. There was, also, some disturbance of the nervous system, expressing itself most frequently in dullness and despondency, hebetude, and inability to sleep refreshingly.

The skin was almost always of a dirty brownish, yellowish hue—a sort of muddy hue. There was, also, in all of the cases, more or less muscular prostration. In some of the worst cases, especially in those where the skin was much undermined, there were colliquative sweats, chills, and saccharine odor of the breath. There seemed to be no tendency to diarrhea. No abnormities in the appearance of the teeth and tongue were observable except in one or two instances. The constitutional symptoms did not seem to have any marked relation to pre-existing constitutional states, either in kind or in violence. Some of the mildest of the constitutional disturbances, the extent of the gangrenous surfaces being kept in view, were observed in those emaciated by previous disease.

10th. The condition of the surrounding tissues seemed to have little or no influence upon the progress of the disease, except in this, perhaps, that the disease advanced rather more slowly but more persistently through those parts solidified by previous inflammation. There was generally to be seen a little tumefaction of the skin along the line of the advancing gangrene, especially where there was much fatty substance underlying the skin. This tumefied skin was often somewhat red, a darkish red; in a few instances, almost livid. This line would vary from the smallest appreciable width to half an inch. In other instances there was no swelling or discoloration of the surrounding edges. The advance of the disease evidently had no connection with any inflammatory process.

11th. The surface of the gangrenous parts was generally insensible. Bromine, iodine, and creosote, produced no pain, unless they reached the living parts below; nor could the gangrenous surfaces be said to be, in any sense, sensitive or irritable.

In some cases the patient complained of a biting, gnawing, stinging pain—a pain always relieved by the efficient application of bromine. In some few cases opium was needed to procure sleep and relieve pain.

12th. Constitutional remedies seemed to have no decided influence over the disease, or over its local or general expression, and had no other perceptible influence than to sustain the strength and relieve pain.

13th. The disease could in no case be said to have a constitutional origin. In no case did the constitutional symptoms precede the local disease, or continue after the gangrene was arrested. It was a noticeable feature in every case, that the constitutional symptoms, whatever their character, as soon as the gangrene was arrested, passed away immediately, and as rapidly as the effect of alcoholic intoxication passes off. In some of the cases there were other wounds in the same individual: thus, one man had a musket ball to pass through the fleshy part of both thighs; in one thigh the wound became gangrenous, and in the other kept on granulating uninterruptedly. In this man's case there was no possibility for the production of the disease by contagion—it was idiopathic. Before the occurrence of any cases in our hospitals, I had directed the surgeons in charge to procure bromine, so as to have it ready for use in case the disease appeared. Many of the surgeons had no experience in the use of the remedy. They were imbued with the idea, prevalent in the profession, that this agent is a highly corrosive and irritating one; and, hence, they almost uniformly used it, in the beginning, largely diluted with alcohol, water, or ether. The inefficiency of this use of the remedy, comparably with stronger solutions, or with the bromine in substance, will be seen in the history of the cases appended. As the surgeons gained experience with the remedy, they gained confidence in its efficacy, and learned that it was not the corrosive and irritating agent which they had supposed it to be. Certain clinical difficulties, however, presented themselves, some of which have already been overcome by further trial, some have not yet been overcome.

It was found that bromine, pure or in concentrated solution, amongst its other effects, coagulated the albumen of the sloughs, thus encrusting the superficial parts so closely as to prevent the bromine flowing into or reaching the deeper parts. The difficulty was overcome, measurably, by first removing the sloughs with forceps and scissors, so that but a thin layer of slough covered the living parts. However, it was not found practicable to remove the undermining sloughs so cleanly. In these cases all that can be done, as yet, consists in pushing the bromine, on pointed sticks of wood, deep into the sloughing cellular planes, or else inserting it by means of the hypodermic syringe through the sound skin into and in advance of the advancing sloughs, or injecting it into cellular planes and gangrenous cavities by means of gutta-percha syringes. What is required in the application of the remedy to arrest hospital gangrene *is to mix the bromine thoroughly with the slough, and touch everywhere the* LIVING *surface of the sore.* This is all.

Later experience with the bromine in my own hands satisfies me that the following directions for its application are sufficient for almost any contingency:

1st. *Preparation.* If the gangrenous sore is large, or the patient intolerant of pain, chloroform or ether should be administered.

2d. The surgeon should be provided with basins, sponges, dry lint, blotting paper if at hand, a stout rubber or a glass urethral syringe, a small measuring glass, some *pure bromine*, a pair of forceps, probe scissors, and a spatula, or the handle of a scalpel.

3d. The patient being prepared, the surgeon should, with forceps and scissors, remove all the sloughs so far that some bleeding points are exposed. The bleeding having ceased, or been arrested by the touch of the bromine, he next scrapes away the fluid putrilage or purulent fluid bathing the surface of the sore; he now turns up the edges of the skin, and, with the handle of the scalpel, removes all the pultaceous matter underlying the skin. The flocculent pulps adherent to the surfaces underneath are now removed with the scissors and forceps. The same proceedings are practised in the cellular planes between muscles. The surfaces are now to be dried, first with lint or tow, and finally with blotting paper, or any dried paper pulp. The bromine is now poured into the glass measure partly filled with water. The syringe, or a pipette, is next thrust through the water into the layer of bromine, and this is drawn up into the syringe or pipette; with this instrument the bromine is applied, first to the cavities between the muscles, next under the skin, next to the exposed surfaces of the sore.

This application of the bromine coagulates and hardens the soft flocculent pulp, and gives the fluid parts of the putrilage the consistence of brain substance. The scissors and scalpel are again put into requisition; the gangrenous portions may now be easily removed, and when it is done, from under the skin, from the intermuscular spaces, and from the exposed surfaces of the sore, the bromine should be re-applied. Where the gangrene attacks the elongated track of a ball-wound—traversing a limb, for example—a piece of candle-wick, threaded in an eyed probe, should be saturated with the bromine and passed through the wound.

Less than this will do for many, for most cases; but this is the effectual method. The beginner with this practice is advised to adopt it closely in his first cases, and as he grows familiar with the use of the bromine he can temper the agent to the necessities of the particular case. Another difficulty was occasionally noticed: the bromine was sometimes applied too often, and thus the separation of the sloughs and the occurrence of granulation were retarded. Bromine is slightly corrosive, that is, applied to a granulating surface, it coagulates the albumen in the granulations and thus kills them; and, to the eye, there is but little difference between a granulating surface which has just been touched with bromine, and a gangrenous surface just touched with it. The rule for its application is this: First, mix the bromine with the slough, in all its parts; then allow the slough to come off. There is no necessity for a second application unless by a return of the fetor it is evident that some part of the gangrenous cavity has not been reached on the first application.

The bromine is sometimes applied too often, sometimes not often enough; sometimes the solution is too weak, never too strong; sometimes not applied thoroughly enough, never too thoroughly. When the bromine touches sound or living abraded skin it gives rise to pain. The pain is sometimes slight, sometimes severe. In a few cases we have had to administer chloroform in order to secure the leisurely and thorough application of the remedy. The pain commonly lasts for a few minutes, sometimes for an hour. The immediate and obvious effects of the application of the bromine to a gangrenous surface are, the coagulation of the albumen and the hardening of the slough. The surface becomes in some parts a whitish yellow in color. *The fetor is immediately arrested.* If any fetor is present after the lapse of ten minutes, or after the odor of the bromine has passed off, the better way is to remove the sloughs still further from under the edges of the skin and apply the bromine; for, just so surely as any fetor remains, there is some portion of the slough with which the bromine has not yet been mixed, and just so surely the disease will not be arrested in all parts of the gangrenous surface. As to subsequent dressings, our convictions are not yet settled: some use dry lint; some, yeast poultices; some, warm water dressings; some, lint, wetted in a weak solution of bromine.

Another important point to which attention is invited is this: No cases treated in these hospitals are isolated; they are treated in the midst of other wounded men. When the bromine is promptly and thoroughly applied, the disease does not spread.

In the beginning, and before the bromine was used promptly and efficiently, a few cases were produced by contagion, but not one after we got into the habit of using the remedy in the way we have now settled upon. In one of the hospitals two cases were brought into the house with the disease full-fledged. A man in the next adjoining bed took it; another directly across the ward took it; another at the opposite end of the ward took it; the weather was cold, the ward was small and ill-ventilated, and all the inmates were wounded men. After the bromine was used, no new cases occurred.

So strongly are our surgeons impressed with this application of bromine, that they have lost all dread of hospital gangrene's spreading in their wards. We have still, every now and then, some new cases, but the most of them are imported. We are continually receiving wounded men from the front; they are brought here in crowded boats and cars, and often their wounds are not dressed from the time of departure to the time of arrival. Most of the cases are doubtless developed *in transitu.**

In the ulcerous and gangrenous affections of the throat occurring during the progress of scarlatina, something of the same effects seem to be be produced by the swallowing of the putrid products of the local disease as are noticed in those having gun shot wounds of the jaw and tongue. Doubtless

---

* While this report is going through the press a number of wounded men have arrived from the battle-field in front of Chattanooga. Some thirty were admitted into Hospital No. 3, in this city. In all of these men the wounds had acquired a gangrenous character. In some, this character was fully developed. They state that their wounds have not been dressed since they left the field, now a week ago. Like events were observed in some of the wounded sent from Philadelphia to Cincinnati. Their wounds were not dressed, as I am informed, during their transit from the one place to the other. In both series of cases the wounds were doubtless in good condition before the men started on their journey. The occurrence of gangrene, in so many instances of a like kind, forces upon me the conviction that the gangrenous quality is produced in these cases by the mere confinement and putrefaction of the discharges of the wounds; for it was noticeable that the men were, in other respects, in good condition.

the putrilage, in the former cases, exercises the usual reaction upon the system at large, noticeable in other gangrenous affections in various parts of the body.

The same indications are measurably to be met in the former as in the latter. The vapor of bromine has been applied in both the scarlatinous and diphtheritic affections, with what effect will appear in the subjoined report. Besides its antiseptic property, bromine seems to have the power of rendering the diphtheritic exudation more brittle, so that the latter is more easily exfoliated, detached, and expelled. I have not observed personally any of the cases of scarlatina, and therefore confine my report on these diseases to the report of cases herewith appended.

The report of Surgeon G. R. Weeks, U. S. V., is appended.

SURGEON WEEKS'S REPORT.

"*Report on Hospital Gangrene, as observed in General Hospitals in Louisville, Ky.*

"I was requested by Surgeon M. Goldsmith, Superintendent of Hospitals in this city, to visit personally all the hospitals in and around Louisville, and collect from the records and Attending Surgeons all the facts that have occurred in connection with cases of hospital gangrene during the winter and spring. Having complied with this request, I beg leave to submit the following statement: I find that, up to the present time, 115 cases have been treated; in 108 of these the gangrene has been arrested, and the patients are either now well or convalescent. Of the whole number treated, 7 have died. 104 were treated with bromine, and of these none have died from gangrene; of this number 80 were treated with the solution of bromine, and 24 with pure bromine. The average time of arrest of the disease in the cases treated with the solution of bromine was 8 days and 19 hours; in the cases treated by bromine undiluted, the average time of arrest was 2.12; in those not treated by bromine, 14.6. Of the number treated by bromine, three have died, two from pyæmia, and one from cellulitis; in all of these, gangrene had been previously arrested, as will be observed by referring to the accompanying report. The condition in each case was verified by a post-mortem examination. In the two individuals who died of pyæmia, metastatic abscesses were found in the lungs, and thrombi in the veins. I have thrown out case No. 13, for the reason that I am satisfied no definite plan of treatment was followed. The facts elicited are these: The man was brought from ——— to Hospital No. 3, in this city; at the time of his admission, his system was much reduced by abscesses of a bad nature, in the region of the thigh, the bone of which was wounded at the battle of Shiloh, on the 7th of April, 1862. The wound became gangrenous after his admission to Hospital No. 3. The surgeon in charge stated that it was impossible to apply bromine thoroughly, on account of immobility of the patient, the position and nature of the wound; and that, in his opinion, it was not applied to all the diseased surfaces. Gangrene burrowed into the cellular spaces; the patient had to be raised to have the bromine applied, and the motion caused so much pain that it was found impracticable to apply the remedy to any useful purpose. I should state also, in this connection, that the patient had a colliquative diarrhea, to which the attending surgeon attributed his death. By reference to accompanying report, it will be seen that eleven cases have been treated by other remedies than bromine. Of this number three died, and nine recovered: of the latter, one was treated with creosote; two by extract of hæmatoxylon; two by chlor. soda, charcoal, and yeast; one by mur. tinct. iron; one by nitric acid; and one by warm water dressings.

"Both these and the cases treated by bromine had the same constitutional treatment, namely: stimulants, tonics, and nutritious diet; with the exception of those treated in Hospital No. 8, where the hyposulphite of soda was used without any apparent benefit; the compound sol. of bromine was also given internally, and the surgeon thought with a good effect. The hyposulphite of soda was given in 10 gr. doses, three times a day, and the solution of bromine was administered in the dose of three drops three times a day.

"It will be seen from the foregoing statements that no deaths occurred from hospital gangrene, directly, where bromine was used; that three of these patients died, subsequently, of other affections; also, that of the eleven cases otherwise treated, three died, or over twenty-five per cent. When the remedy was first introduced at this post, by Surgeon M. Goldsmith, it was used mainly in the strength of bromine 1 part, water 24. Under this treatment it will be observed, that the average date of arrest was 12.61 days; but after Smith's formula was prescribed the average duration was 8.19; and that in the twenty-four cases treated by pure bromine, (these were the worst cases,) the average was 2.12 days—showing, conclusively, that the pure bromine arrested it in the shortest period of time. In considering the several interesting facts falling under my observation, the following seemed peculiarly note-worthy:

"1st. The strength of the article used.
"2d. The condition of the parts to which it was applied.
"3d. The manner of its application.

"The solution of the varying effects of the treatment seems to me to be found in the study of the foregoing points.

"I found a common error in the cases where the bromine had appeared to act inefficiently. Sometimes the article used was not of sufficient strength; in others, the wound was not well prepared to receive it; and in some, it was applied too often, thus preventing the natural powers from establishing the reparative process in the wound, even after the complete destruction of the *materies morbi* in the part. From all the facts observed, I believe the following mode of application generally to be the best: First, the sloughs should be cleanly dissected away until we meet evidences of vitality, or by hemorrhage are warned to go no further—being careful to clean out all nooks and corners where the gangrene has dipped down into intermuscular spaces, or followed along cellular planes, where, unobserved, it makes its nidus safe from observation, too surely to resume its onward progress if not reached by the remedy.

Second: After having removed all the dead tissues, the wound should be washed thoroughly in warm water, injecting with an ear syringe all the parts otherwise difficult to reach. Then the wound should be effectually dried with a pledget of lint or charpie, using the same care in regard to small cavities extending from the central sore; for the reason that if bromine comes in contact with the albumen, or solution of albumen, it immediately coagulates it, and thus forms a barrier to the further diffusion of the remedy. I saw many evidences of the violation of this rule whilst visiting the hospitals. The surgeon would frequently remark: 'The gangrene was arrested everywhere, except in two or three small points as large as the end of the finger;' and these were always found to be rich in cellular tissue, occupying intermuscular spaces, along which the process had traveled with more rapidity than at any other, and. consequently, it frequently extended beyond the reach of the remedy, and passed beyond the limit of vision. Third: Having the wound thus prepared, pure bromine should be applied with a mop or swab to every part of it. If all the surfaces cannot be reached in this way, the bromine should be injected into the smaller cavities with a glass syringe, (and with a pointed stick of wood, adapted to the dimensions of the cavity, with the end covered or not, as the case may require, by a piece of soft cloth or lint.) It should be pressed up and thoroughly mixed with all the pulp or pultaceous fluid that may still linger in the wound. Simple dressings should then be applied, and the wound excluded from the atmosphere. On the second or third day after the application, warm water dressings should be used to facilitate the detachment of the sloughs, and to wash away all molecular matter that may have accumulated in the wound. In four or five days granulations may be observed springing up, and, if no fetor be present, the cure will be complete. However, if some fetor still lingers about the wound, it is evident that some points have escaped, and these should be re-touched, observing the same rules as in the first application of the remedy, and being careful to apply it only to the places where the gangrene is not arrested. The wound should then be treated on general principles. The only case where I would use the *solution* of bromine is, as a local stimulant, where the granulations are weak. I believe, where pure bromine is thus applied, one application is sufficient to arrest the disease. As yet, no case has occurred which has needed a second application of the remedy where it has been used in the manner heretofore stated; and, for my part, I cannot imagine the condition that would demand it. In regard to the repetition of the application, I consider it of as much importance to await, after one efficient application, the detachment of the sloughs and commencement of granulation, as I do the application of the remedy in the first instance. It is just as impossible for repair to be set up in a wound where bromine is used frequently, as if nitric acid was applied in its stead. Although the process is stayed, repair is prevented by the frequent re-application of the remedy. This fact is well illustrated in Case 93, where pure bromine was said to have been applied twice a day for a month before granulations made their appearance, and the surgeon might have added that they would have never been observed if he had not ceased to apply the remedy; for then granulations made their appearance, which would have been also the case if he had ceased the application sooner. This is the explanation of all the cases reported in which bromine was used for the period of one or two weeks, and was reported to have failed to arrest the disease. When bromine was first used here, it was commonly applied in a weak solution three times a day. It will be seen that, in some cases, after the remedy was discontinued, or another of a more harmless character substituted, the wound rapidly granulated. As a case in point I refer to Case 25, occurring in Hospital No. 4. In this case the disease resisted the action of the compound solution of bromine, as the surgeon stated, and was arrested by a weak solution of creosote.

"It will be seen by noticing the three following cases (as the surgeon treated them all in the same manner) that he was of the opinion that in creosote he had a better remedy than bromine—a proposition he has failed to verify, as will be learned from Case 30, where creosote was used for ten days without any perceptible arrest of the process, after which it was arrested by three applications of bromine. This case I saw. It was on the inner aspect of the left thigh, below Scarpa's triangle, and was about six inches in diameter. The same length of the internal saphenous vein, running directly across the sore, was destroyed. This wound is now rapidly filling up with granulations, and the patient doing well in every respect. In the investigations of these cases, many interesting facts were observed in regard to the effect of the local application of bromine to gangrenous sores on the constitutional symptoms, the entire weight of evidence being that the constitutional disturbance and manifestations began to disappear as soon as the local disease was arrested by the remedy. This was observed by friends and visitors, as well as physicians, namely: The rapidity with which the constitution rebounded after being relieved of its burden. The anxious and pinched expression of the countenance, the general feeling of lassitude, languor, and debility, the leaden hue of the surface, rapidly passed away; the skin, if clammy and bathed in perspiration, became warm and dry; the pulse rose in strength, and decreased in frequency; the appetite returned; and, in short, all the organs began to assume their natural tone and vigor. These effects were observed very constantly, immediately after the first application of bromine, and could not be attributed to the constitutional treatment, as the same condition was observed where no constitutional treatment had been used. I leave the argument for others; my limits will only permit the simple record of the facts.

"I wish also to direct attention to another very important point, namely: the practicability, by the aid of bromine, of tying an artery in a gangrenous sore where secondary hemorrhage has occurred.

"If bromine will arrest gangrene as certainly and as speedily as this report shows, can there be any objection to tying the artery in the wound? thus following the original rule of Guthric, directing that the bleeding vessel, if it can be reached, shall always be secured in the wound. If we tie it immediately beyond the dead and in the living tissue, and if by the aid of bromine we are certain of arresting the further invasion of the parts by the gangrenous process, what objection could be urged against this procedure? It is in cases of this character that bromine holds superiority over nitric acid and all other known remedies—the capability of arresting the gangrene without destroying important parts which cannot be avoided in the application. Case No. 4 presented itself, and I embraced the opportunity of ascertaining the facts from actual observation.

"I was requested to see this man, and found he had had secondary hemorrhage to the extent of about fifty ounces. I found him pulseless and cold from loss of blood. His leg had been amputated January 1st, 1863. The flaps had all sloughed off, the tibia projecting two inches. I directed that he should have stimulants, and waited a short time until he partially rallied. I then prepared the wound as heretofore directed. Upon loosening the tourniquet, the anterior tibial artery bled freely; I tied it in the wound just below the junction of the upper and middle third of the tibia, and applied pure bromine effectually to the wounded surface. I kept the tourniquet on loosely, and an attendant, instructed in its use, was constantly at the bedside, day and night. On visiting the patient next day, I found him very comfortable; pulse 100; skin warm; appetite better; looking bright; anguished countenance displaced by a more cheerful one. There was no fetor emitted from the wound, and the patient was every way doing well; the ligature came away on the fifth day—no hemorrhage appeared; he is still convalescing; wound is filling up slowly. In this case granulations were slow in making their appearance—attributable to the great loss of blood. Applied weak solution of bromine to the wound; the case is now convalescing rapidly, and no hemorrhage has occurred. Two other cases illustrating the same principle, to some extent, have occurred in hospitals in this city. One at Hospital No. 3, Case 6 of accompanying report, where the brachial artery was tied in a gangrenous sore with a like result. Another occurred at Hospital No. 12, Case No. 77; where the dorsalis pedis artery was tied under the same circumstances and with the same result.

"After having visited all the hospitals, gleaned all the facts, heard all the evidence, and conversed with all the surgeons, the impression left upon my mind is, that in bromine we have a remedy *certain* in its effects for the arrest of hospital gangrene, the greatest scourge of military hospitals. This I am aware is strong language, but I think not more so than the circumstances and evidence in the case warrant. I expect most confidently the future will verify what I am now saying. Bromine has robbed gangrene of its terror, and shorn it of its power to stalk through the wards where the sick and wounded are congregated, spreading its contagious and pestilential influence in every direction. But, armed as the surgeon now is by the use of a remedy so certain in its effects, a feeling of security pervades the entire profession at this post, not only in this, but in all that family of diseases formerly supposed to have their origin in blood poisoning, namely: erysipelas, gangrenous diphtheria, scarlatina, &c.; its use allows us to class hospital gangrene among those diseases over which the surgeon has as absolute and complete control as the physician has over intermittent fever. I am forced to the conclusion, from all the facts presented, looking at them, I think, with an impartial eye, that bromine will as surely arrest hospital gangrene as quinia will ague.

"GEO. R. WEEKS, *Surgeon U. S. V.*"

A few remarks are appended which force themselves so strongly upon my attention that I feel I must give them utterance, however heretical and revolutionary they seem in the presence of traditional dogmata. It will be seen that in the history of the several affections treated of in this report, one fact seems to pervade all that is said, and that is, the reactive influence of the local pathological processes—the constitutional contaminations caused by the absorption of the fluids produced in the local processes.

That constitutional contaminations, general disease, may be produced by local maladies, is shown by the occurrences observed in syphilis, in vaccinia, in glanders, in rabies canina, and, from what is recorded in the cases hereto appended, in hospital gangrene. In these diseases it is evident that the abortion of the local processes prevents constitutional contamination; or, if such contamination exists, as in hospital gangrene, the symptoms peculiar to the general malady disappear quickly after the arrest of the local process.

Virchow, I think, has established the general law that in the sense commonly employed there is no such thing as a permanent dyscrasia. He has shown that the syphilitic dyscrasia is but a continuous infection arising from the continued production of syphilitic matter in the lymphatic glands of the neck;—that with the subsidence of the lymphatic enlargements, the syphilitic dyscrasia disappears;—that the cancerous cachexia is but a continuous infection from the diffusion of the cancerous juices. The vaccine infection disappears with the cicatrix—this is a common impression. When the surgeon examines the arm of a person, to see if re-vaccination is needed, if the characteristic scar has disappeared, he re-vaccinates; if he finds a scar good, he does not re-vaccinate. Besides, there is good reason to believe, though the cases have not been numerous enough to warrant the statement as a settled fact, that those who have lost the vaccinated arm by amputation at the shoulder joint, lose their protection; so that we are at least warranted in entertaining the suspicion that the vaccine dyscrasia, so to speak, is a continuous infection from the cicatrix.

Further: we are prepared to believe that this is true of the foregoing diseases, because we see in them a regular series of events constantly recurring in the order and relation of cause and effect. If we could see a certain set of constitutional disturbances, tolerably constant to certain local processes, (as, for example, the symptomatic fever of small-pox, or the waxen face of scarlatinous fever coincident with ulceration of the tonsils,) disappear whenever the local processes were arrested, surely we would be authorized, whatever our idea of the original disorder may have been, to believe that, in the particular states, the symptoms referred to were due to some influence exerted by the local processes themselves.

Now what are the facts?

1st. If the small-pox pustules or vesicles are opened, and if iodine, bromine, or nitric acid is applied, the vesicle or pustule is aborted; a scab is immediately formed, it soon falls, and if the whole crop or the major part of the crop of pustules has been aborted, the irritative fever is also aborted—the whole constitutional disturbance abates, and immediate convalescence ensues.

2d. If the ulcerous surfaces of the tonsils, in scarlatina, are brought into a healthy condition—nay, more, if the peculiar processes are arrested—nay, still more, if the fluid products of the ulcers are not swallowed, and are rendered inert by the application of any re-agent which breaks up their chemical structure—the constitutional state passes off.

3d. The same is true of diphtheria. In regard to the two last mentioned diseases, reference is made to the individual observations of the reader. Is it not within the latter's knowledge that groping empiricism claims no blood or constitutional antidote, specific and peculiar to either of the blood states, or to both? She supports the strength, just as she does in typhus, pyæmia, erysipelas, and other disorders of "a low type." The great reliance of medicine is in topical remedies, the mineral acids, chlor. ferri, chlor. sodæ, &c., &c. Now why? Surely if the local state is but the local expression, or a part of the local expression, of the general state (for so the saying goes), what need of addressing ourselves to the mere outcrop of the disease, while the real gravamen is in the blood behind all this? Is it not, rather, that empiricism, a sounder teacher than a crude philosophy, has somehow felt that the constitutional state is in some way dependent upon the local state? Whether this view is frankly avowed or not, the idea stands confessed in the treatment instituted; for if the local treatment does not embrace this idea, then it is blind and purposeless indeed. True, sometimes the local processes in both scarlatina and diphtheria produce death without reference to the constitutional state. But how often does this occur, in comparison with the frequency of death through the increasing gravity of the constitutional symptoms?

4th. Pyæmia is the constitutional state attending upon, or rather occasionally produced by wounds—*e. g.*, the theory runs that patients by breathing a poisoned atmosphere acquire a constitutional or blood state, eventuating in if not consisting of pyæmia. Now it is a singular circumstance to begin with, that no one has pyæmia who has not, somewhere in his body, infiltrating tissues, filling cavities, or flowing from wounds or their equivalents, animal fluids whether regarded as exudates without morphological change, or in the form of cell-bearing liquors. Again, it is within the knowledge of every medical man that pyæmia may occur independent of suppuration, in connection with pathological processes where the warmest of Virchow's admirers fail to discover any cell-growth or cell-differentiation; nay, more, pyæmia is more constant to such collections as are found in exquisite erysipelas, where the purulent fluids yield to the most searching investigation neither cells, nor nuclei, nor nucleoli—nay, not even a connective tissue, corpuscle, or any morphological structure; and that seemingly, just in the degree that the morbid probid products approach the standard of true pus the danger of pyæmia grows less. Does any surgeon anticipate pyæmia in a so-called healthy suppurating wound? Does he not, rather, when healthy suppuration takes place, fetch the long breath that betokens a danger passed—an anxiety relieved? I appeal to universal experience.

Yet the co-existence of pyæmia with the conditions mentioned above is so constant and invariable, that the common instinct of medicine associates them in some sort of relation of cause and effect.

More than this: the every day record of cases shows the co-existence of pyæmia rather with ill-conditioned discharges; and every day's record, also, sets forth the amendment on the opening of confined collections, the use of drainage tubes, the cleansing of the parts, &c., &c. It may not be and often is not stated that the amendment was because of these events, but the coincidence occurs. Still more: the appended cases, noticed with peculiar reference to this matter, show that this amendment coincides (with a certainty that is remarkable, and a rapidity that is marvelous) with the injection into the infecting cavities of substances which, by strong coercion, arrest the putrefactive motion, destroy the products of putrefaction, and render putrescible substances non-putrescible.

5th. The constitutional condition allied to hospital gangrene is a sequent state. The proof of this statement is abundant in the record of cases hereto appended. The reader's attention is invited to the constant subsidence of the general or constitutional disturbances upon the arrest of the gangrene, a subsidence constant, marked, and immediate—most emphatically pronounced in those cases where the bromine was used undiluted, and carefully applied. The attention of the reader is called to the ameliorations corresponding with the amelioration of the gangrene, to the constant persistence of some general expression so long as any part of the surface was untouched or unchanged. So constant was the loss of appetite, debility, etc., to the lingering of the gangrenous process in limited portions of the surface, that the very hue and expression of the face betokened the fact, told the story to the observant surgeon. It is needless, for my present purpose, to multiply examples. I desire, here, merely to convince the reader that, in some diseases, THE GRAVAMEN OF THE CONSTITUTIONAL STATE, IF NOT ITS TOTALITY, IS PLAINLY DUE TO THE ABSORPTION OF THE PRODUCTS OF THE LOCAL PROCESSES. I say, *absorption;* for it is not possible to conceive of any other process by which the whole organism could be involved to the extent noticed.

Now, if the proposition is proven in regard to the diseases in question, it may reasonably be asked if it is not also true of other disorders—if it is not, even, within certain bounds, a law in medicine? I confess that my observations lead me to think it is one of broader significance and wider application than is generally believed. The matter, of course, needs a full investigation. I do not desire, in what is here written, so much to challenge belief as to invite investigation. If the projected law is a true one, the effect would not be to revolutionize but to simplify; to give precision to methods now vaguely used; to give definite views and purposes to remedial measures; to draw attention to the completeness in the effect of traditional remedies; to supplant surmise with faith, and indecision and doubt with confidence.

CASES.

[The following cases are selected in order to show, amongst other things, the varying effects of the different modifications in the use of bromine, and to give the reader some idea of the general development of the clinical processes elsewhere recommended.]

Wilbur F. Nichols,[*] corporal company B, 34th Illinois, was admitted into Ward No. 1, Hospital No. 7, January 15th, 1863, having a flesh wound upon the inner aspect of the left leg, in its lower third. The wound was made by a minie ball, had its aperture of entrance separated from that of exit by a piece of integument about two inches in width. The wound was superficial. The edges had a contused and purple appearance.

January 20th, well marked hospital gangrene made its appearance. The first application made was lint saturated with liq. chlor. soda. This application was continued for some days, with slight improvement in the cleanliness of the wound. The latter was covered with large, greyish, pulpy sloughs, and a

[*] Nichols's case was the first one of hospital gangrene to which the bromine was applied.

scrofulous looking pus, emitting a foul odor; it is about five inches in length in its largest diameter. The tendons are to be plainly seen, as well as the internal saphena, exposed, indurated, filled with coagulated blood, and sloughing throughout the whole extent of the ulcer. The tibia and fibula are bare, the former for two and a-half inches. Constitutional symptoms are prostration, restlessness, sleeplessness, and loss of appetite. The local treatment has been, up to date February 10th, mur. tinct. ferri, charcoal, and cinchona poultices, tar water, and dilute nitric acid. Constitutional treatment has consisted of wine, quinine, egg-nog, and generous diet. This morning, on removing the dressings, an arterial jet followed, which was stopped by torsion. There seems now nothing left but amputation. Although the patient is willing to submit, the operation holds out but little hope, on account of the great prostration present.

Surgeon Goldsmith, U. S. V., Medical Director, was called upon this morning, in order to obtain his consent to the operation. He ordered the treatment by bromine, to be commenced by injecting it into the wound in all its parts, particularly under the raised edges of the skin. In the meanwhile the sloughing had extended in the cellular tissue from the ankle nearly to the knee. The foot was hugely swollen. The wound is covered with thin, diffluent, greyish, fetid sloughs of skin, cellular tissue, and muscle. A dirty yellowish fluid oozes from the cellular planes. The tibia is bare about three inches.

    ℞—Bromine, gtt. xx.
      Alcohol,   ʒj.
      M.

To be applied every four hours, day and night.

February 11th. The wound looks cleaner this morning than has been seen for a long time. Sloughs not so adherent, ulcer is becoming more sensitive; continue treatment.

February 12th. The improvement in the wound is marked: patient feels better in every way; appetite and spirits better.

R—Bromine, gtt. xl.
Alcohol, ʒj.
M. Apply as before.

February 13th. Whole ulcer presenting a healthy granulating surface; appetite and all constitutional symptoms very much improved. The bromine was now suspended on account of the pain. The wound has continued to granulate until, at the present date, March 7th, it is on a level with the surface. Patient is daily gaining in strength, and often speaks of the substance which he thinks saved his leg and life.

[Condensed from the report of Medical Cadet Larabee.]

Nelson Koroson, private company B, 10th Wisconsin Volunteers, was admitted into Ward 1, Hospital No. 7, Louisville, Ky., January, 15th, 1863, at 9 A. M., having, in battle at Murfreesboro, Tennessee, received a flesh wound upon the inner aspect of the lower third of the right thigh—apertures of exit and entrance near together. The wound has a contused and purple appearance, as have the tissues for several inches around. General health good.

January 20th. Gangrene made its appearance February 17th. The wound is now four inches long, three inches wide, and one and a half inches deep. The topical applications have been warm water dressings, sol. acet. zinc, tr. ferri chlor., tr. arnicæ. Patient has had frequent chills, diminished appetite; has taken cathartics, quinine, Dover's powder, and generous diet. Large sloughs have been removed every day. The skin is undermined to the extent of two inches in every direction. Commenced to-day to inject the ulcer with tinct. of bromine, as in the case of Nichols, and in the meanwhile bromine vapor was applied in the usual manner.

February 18th. Morning visit. The sloughs are of a yellowish color and easily detached; wound is now sensitive, and shows more signs of vitality than have been seen before. Continue bromine. Bromine causes pain.

February 19th. This morning patient says he feels better than he has for a long time; wants something good to eat. Sloughs have all disappeared from the ulcer; a healthy, laudable pus is secreted; granulation may be seen commencing, foul odor gone; whole surface a lively red color. From this date the recovery was rapid.

It is proper to state that in all the cases coming under my observation in which bromine has been used, simultaneously with the removal of the sloughs, by the arrest of the gangrene, the appetite and general health of the patient has improved.

[Condensed report by Larabee, Medical Cadet.]

J. W. Bennet, private, company H, 44th Tenn. infantry, C. S. A.: wound inflicted by a fragment of a shell at the battle of Murfreesboro, December 31st, 1862. The inner aspect of the middle third of left thigh torn away, with much contusion of the surrounding tissues. Sloughing and gangrene continued until the ulcer covered a space of eight inches, longest diameter, and five inches, shortest diameter, the tissues being destroyed an inch in depth. The fleshy part of the lower border of the rectus femoris, the belly of the sartorius, the adductor magnus were involved in the sloughing, and the long internal saphenous vein was exposed throughout the extent of the ulcer until it sloughed away. Admitted for treatment Feb. 15th, 1863. Age 27, a farmer, height 5 feet 6 inches, born in Middle Tennessee; complexion dark; for the last six years suffered greatly with his breast; temperament, nervous. Has irritable stomach, furred tongue, fever, constipated bowels, with great restlessness. Ordered saline draughts until bowels operated; wound dressed with flaxseed poultice; sulph. morphiæ, 1 gr., at bed-time.

February 16th. Stomach still irritable; tongue moist but pale, fur cleared off; bowels moved twice during the night; slept well and feels refreshed; wound looks and smells horridly. Great quantities of slough removed by poultice and scissors. Dressed with lint saturated in bromine, 1 oz., alcohol five oz. Also the above solution injected into and all over the ulcer without producing any pain. Nothing but milk and lime water retained in stomach. Egg-nog, milk punch, whisky, and brandy toddy rejected, as also sherry wine.

17th. Had slept well after taking 1 gr. morph. the night previous; stomach still irritable, retaining little but milk and lime water, with some cold chicken broth; ulcer looking and smelling slightly better; bowels moved once; some fever still. Bromine treatment continued to ulcer, the dressings enveloped in oiled silk. Saline draughts continued at intervals of three hours. Barley water (to satisfy intense thirst) in small quantities. Sinapisms over epigastrium.

18th. Stomach less irritable, fever slight, ulcer decidedly improving. Bowels moved once. Bromine treatment continued. Sulph. morph. one-half grain every four hours; light diet, with corn-bread, allowed.

19th. All symptoms improved; ulcer tolerable in center, deep sloughs around the edges. Bromine treatment continued. Light diet, with corn-bread and Sherry wine, retained and continued. The patient feels better, this being the first time he has so expressed himself.

20th. Sloughs still adhere, and their removal produces nausea without pain. Bromine produces slight tinglings, but no pain. Continue treatment.

21st. No change.

22d. Sloughs still deep, but their removal produces but slight nausea with considerable pain. Continue treatment.

23d. The ulcer very painful. Indisposition on the part of patient to have it dressed. Sloughs around the edges deep and ugly; fever, irritable stomach returned, with loss of appetite and profuse perspiration. Bromine treatment continued. Tinct. chlor. ferri, gtt. xx., every four hours, sulph. morph. ss. grain every 3 hours. Sinapism over the epigastric region. Chloroform administered whilst applying the bromine.

24th. No change except stomach less irritable, perspiration not so profuse. Our Medical Director, Dr. Goldsmith, on his visit to-day, was kind enough to dissect off the sloughs, and apply his solution (bromine ten parts, water ten parts, bromide of potassium two and a half parts), with a wooden spatula, directly to the parts that required it. Light diet and wine was continued.

25th. Sloughs and softening disappearing. Continue treatment.

26th. Still improving, only two or three points to be touched with the bromine. Appetite good, stomach retentive, perspiration natural, tongue clean, fever absent; takes egg-nog freely, chicken soup, baked apples, corn-bread, &c.

27th. Most beautifully filling up with the nicest granulations, no points visible for the action of the bromine. Simple cerate dressings ordered. From this time the recovery was uninterrupted, rapid, and complete.

A. T. C. WORTHINGTON, A. A. S., U. S. A.

J. B. Mortimer, co. B., 5th Arkansas, C. S. A., wounded by a conical ball, December 31st, 1862, at the battle of Stone River, Tennessee, was admitted February 15th, 1863; age twenty-four years, height five feet four and a half inches, complexion fair, strumous diathesis. The ball entered above tuberosity of ischium, left thigh, and passed out and over the great trochanter. The point of exit, larger than the point of entrance, was separated from the latter by a band of integument over the track of the ball; this continued to slough until the two apertures were united in one large ulcer, upwards of five inches in diameter. The sloughing involved the glutei muscles and the adipose tissue. The trochanter major was denuded of its periosteum. The patient exhibited the following symptoms on

Feb. 15th. Delirium, with well marked characters of hectic. The wound was dressed with the usual water dressings, with the internal use of saline draughts, cold water to the head, &c.

16th. Not much better. Liberal diet, stimulants, viz: egg-nog and milk punch, in pint quantities, daily. Warm water dressings.

17th, 18th, and 19th. Treatment continued.

20th. Wound looking very bad. Bromine ʒj., alcohol ʒiv., applied every two hours by saturating lint after well injecting the whole surface with the above solution. Liberal diet.

21st. Improvement. Occasional delirium, fever not so intense, wound looking improved in the center, sloughing going on around the edges. Treatment continued.

22d. Treatment continued.

23d. Dissected all the old sloughs away, and applied the comp. sol. bromine, at the suggestion of Dr. Goldsmith, with a wooden spatula, directly to all the parts having a sloughing tendency.

24th. The greatest improvement imaginable, viz: no delirium, no fever, appetite good, is cheerful, and feels like getting well at once; wound granulating finely; everything going on well. Treatment continued.

25th. Still improving. Treatment, warm water dressings. Wound filled with fumes of bromine.

26th. Still doing well. Same treatment continued.

27th. Wound nearly filled up with the granulations; general health very good. From this date convalescence progressed rapidly. The exposed bone granulated freely.

A. T. C. WORTHINGTON,
A. A. S., U. S. A.

Milo F. Baxter, sergeant, company F, 11th Kentucky, in good health, was wounded in the battle of Murfreesboro, January 3d, 1863, the ball passing through the fleshy part of the left arm, about three inches above the elbow, leaving a space of about two inches between the point of entrance and exit. He was placed at convalescent barracks, where the wound had received but little attention. Admitted February 15th. Condition when admitted: Patient feeble, tongue coated and dry, bowels constipated, pulse 115, appetite not good, restless at night.

The wound presented two ulcers, about an inch in diameter and an inch apart, covered with greyish, thick, and rather firm sloughs. Twenty-four hours after, the intervening space had been destroyed by the disease, and the two ulcers became one, then about three and a half inches in length and two and a half wide, covered with the same firm, greyish matter, and exceedingly offensive. The edges of the ulcer were red and tender. The patient suffered great pain. Treatment and progress: for three days the ulcer was treated with warm water dressings, and continued to extend. On the morning of the 19th bromine was first applied—about one drachm to a pint and a half of water—with but little or no improvement. The ulcer continued to extend, burrowing under the skin, and the thick sloughs still remaining. On the 21st the sloughing invaded the brachial artery, causing profuse and almost fatal hemorrhage. The hemorrhage was arrested only by tying the vessel, which was done about an inch above the border of the ulcer. By the 23d, two days after the artery was tied, the sloughing had extended to that wound also, and great fear was entertained that hemorrhage might again occur. At this time the ulcer had attained the size of about five inches in length, and three and a half in width, the patient very feeble, pulse 140 per minute, tongue dry and coated with a dark brown coat. It was then resolved to try a strong solution of bromine, applied to the sloughs, as well as to the brachial artery which had been tied, and to the parts adjacent, the whole ulcer to be covered with lint saturated with a solution of one part of the comp. sol. bromine to sixteen of water. This dressing was repeated every four hours, good diet and ale were prescribed, and an opiate at night. Under this treatment the patient improved rapidly, the slough separating, leaving the surface beneath granulating. The ligature held, and during the granulating process the tied extremity of the brachial could be seen in situ, covered, like the other tissues, with granulations. The bromine was diluted as the sloughing ceased and the parts became more sensitive. On the 5th of March the ulcer presented a healthy appearance, entirely free from gangrenous matter, filling up rapidly with granulations, and already contracting in size.

March 12th. Improvement continues without interruption.

The character of the ulcer, the rapidity with which it extended, the great constitutional disturbance, and especially the exhausting hemorrhage rendering it necessary to tie the artery, all combined to make this an interesting and instructive case. From a somewhat close observation of this case (it was not directly under my charge) I feel assured that the sloughing might have been arrested sooner, but for some fault in applying the bromine. It was not well understood then, if indeed it is yet, what is the best mode of applying the remedy, and in what strength; and I am satisfied that it was not efficiently applied in the earlier part of the case. The solution used was not strong enough, and the dressing was not applied often enough.

It is worthy of notice how soon the general condition of the patient improved after the local disease was arrested. This could not, I think, be attributed to the general treatment, although that unquestionably had a share in the improvement. The rapid improvement in the appearance, appetite, pulse, tongue, and nervous condition of the patient, so soon after the spread of the gangrene was arrested, could not be mistaken for the cause of the improvement in the ulcer.

J. A. DOUGHERTY, A. A. S.,
March 12th, 1863. *Hospital No. 3.*

Henry Herman, private, company A, 6th Ohio, previously in good health, received a flesh wound (gun shot), December 31st, 1862, at Stone river, Tenn., the ball entering to the left and lower extremity of the sacrum, passing downwards and outward, and emerging about five inches from the point of entrance. The wound did well for about four weeks, when erysipelas set in, attended with considerable tumefaction, extending to the left groin, and resulting in hospital gangrene. The patient was admitted to Hospital No. 3 about ten days after gangrene set in, viz: February 15th. Condition when admitted: Patient considerably emaciated, feeble, pulse 110 per minute and feeble, appetite not good, tongue coated and rather dry, bowels constipated, restless at night.

In the left groin, just above Poupart's ligament, was an ulcer about six inches in length, and four and a half inches wide, and at least three-fourths the thickness of the abdominal walls in depth. Sloughing was still going on, the ulceration burrowing under the border of the skin, which was of a dark, dead appearance. The surface of the ulcer was covered with a dirty greyish slough, and exceedingly offensive. The sloughing had exposed one of the inguinal glands, which stood out from the surface as large as a partridge egg.

Treatment. For two days the ulcer was treated with warm water. On the 17th applied bromine, in weak solution, (brom. 3ss., ether 3v.), by saturating a piece of patent lint with the solution, and, after first covering the ulcer with dry lint, laying it on, then enveloping the whole with oiled silk to prevent the escape of the vapor.

On the 18th the bromide of potassium was obtained, and the comp. sol. was prepared and used in the subsequent dressing. This solution, diluted more or less as the patient could bear without great pain, generally about one part to twenty-four of water, was applied, as a wash, by means of saturated pieces of lint, as often as every four hours until the twenty-sixth, about nine days. Good diet and ale were prescribed, with aperients as often as the condition of the bowels required.

For the first twenty-four hours, whilst the etherial solution was being used, there was little or no improvement. After the comp. sol. was resorted to, the improvement was well marked in six hours, and continued without interruption until at present.

March 9th. The ulcer is filled up with healthy granulations, and has contracted to less than half its original size. On the 3d of March a deep erysipelatous redness was observed, extending from the border of the ulcer backwards. This was dressed with a weak solution of bromine—one part of the comp. sol. to sixty of water. In twenty-four hours it had disappeared. This was the only deviation from a uniform recovery in the case.

J. A. DOUGHERTY, A. A. S., *Hospital No. 3.*
March 12th, 1863.

Linnieus O. Smith, aged about 31 years, was wounded at the battle of Murfreesboro, December 31st, 1862, by a musket ball entering the inner and upper part of the left leg, passing obliquely downwards and outward, traversing the gastrocnemii muscles, and emerging about six inches from the point of entrance, having slightly grazed the tibia, but left the principal vessels intact. Was admitted to General Hospital No. 8, Louisville, Kentucky, January 14th, 1863, with much fever, furred tongue, deranged secretions generally. The wounded parts much swollen, very painful, and discharging a sanious and offensive pus. A few days subsequently the parts became gangrenous, involving the cellular tissue, and forming sinuses in the gastrocnemii, covered with dark, thick, adherent slough, the stench of which it was almost impossible to remove from the ward.

The treatment had consisted of simple water dressings—with the internal exhibition of quinine, iron, alteratives, and a nourishing diet; but the gangrene remained unchecked nevertheless.

Surgeon M. Goldsmith, Medical Director, now ordered an application of bromine to be made; and having removed the dark, fetid, and adherent sloughs, with scalpel and forceps, *pure* bromine was applied to the surface, giving much smarting for some fifteen or twenty minutes. An emollient poultice was applied, and, on its removal a few hours after, a very sensible change was observed in the character of the discharge, and especially in the correction of the offensive odor. Twenty-four hours after the first application of the bromine a second application was made to the parts, this time Smith's sol. This was injected freely into the sinuses, and, after a few moments, pressed out again gently with the hands, and the poultice applied as before.

In forty-eight hours after the first application of bromine, the gangrene was entirely arrested, all offensive, putrid odor removed, and the parts discharging a much improved character of pus. A weaker solution of bromine, ʒss. water ʒviii., was applied daily as a wash for three days, and the parts were dressed with ung. resin. Tonics and nourishing diet, with wine, ale, beef tea, &c., were continued as before. A very marked change in the general health and appearane of the patient occurred soon after the first application of the bromine. The night sweats and the diarrhea, previously existing, ceased directly, tongue cleared off, appetite increased, and general expression and appearance improved. The wound is now, February 1st, covered with healthy granulations, and is rapidly closing. His general health is good, and he is doing well in every respect. This was a bad and very unpromising case of hospital gangrene; and the marked and rapid change in the arrest of the gangrene, the correction of the fetor, and the general improvement of the patient, were particularly gratifying and instructive.

[Signed,] FRANKLIN IRISH,
Surgeon 77th Pa. Vols.,
In charge of Hospital No. 8, Louisville, Ky.

Wm. Murphy, of co. I, 78th Pennsylvania Regiment, aged 24 years, entered this Hospital on the 28th of February, 1863. He was wounded at the battle of Murfreesboro with a conical ball, passing through the right foot, fracturing the metatarsal bones of the second, third, and fourth toes, portions of which bones had been removed at Murfreesboro. On entering this hospital his wound was in a healthy condition, his general health was good, and remained so, until the 15th of March, when the wound began to present a gangrenous appearance, being of a dark, purplish, grey color, and of an offensive odor. On examining the pulse I found it beating one hundred and twenty to the minute. Skin hot and dry, tongue furred, breath of a saccharine odor, bowels constipated, no appetite, and an aversion to animal diet.

Made an application of the solution of bromine, full strength, (Wilson & Peter's solution), by means of hair pencil, three times a day, and then applied cinchona poultice. Gave

℞—Chlorate of potash,
 Calomel, aa. grs. x.
 Ipecac pulv., grs. x.
 Opii " grs. ij. ss.
Divid. in chart x. One every three hours.
Also—
℞—Quiniæ sulph.,
 Ferri sulph., aa. gr. xxiv.
Divid. in ch. xii. One before each meal. Extra diet.

March 16th. Great improvement in the appearance of the wound, pulse regular, skin moist, tongue cleaning off, appetite improving, bowels open. Continue the application of bromine, and cinchona poultice. Quinine and iron, as on the day before. Continued this treatment for three days, at the end of which time the wound presented a healthy character, and the general health was entirely restored.

J. S. LOGAN, A. A. S.,
General Hospital No. 12, Louisville, Ky.

Ery W. Taylor, company C, 19th Michigan, wounded at the battle of Spring Hill by a musket ball, which entered on the left side and anterior portion of the neck, passing under the sterno-cleido-mastoid muscle between the external jugular vein and carotid artery, making its exit near the anterior border of the trapezius muscle, four inches from its attachment to the occipital protuberance. He was admitted to General Hospital No. 11, Louisville, Kentucky, April 15th, forty days after the reception of the wound. The point of entrance had entirely healed, but that of its exit was a deep, indolent ulcer, smooth, clean edges, about the size and shape of a large almond. His constitutional condition was far below par, yet he was able to walk about and go to the table for food. He was ordered a nutritious diet, with wine, and the wound dressed with citrine ointment.

April 19th. Wound dark and indolent, and the odor fetid; pulse 100, tongue dry with a thin white coat in the center and red edges, some diarrhea, a severe cough; sleeps poorly and has night sweats; a charcoal and yeast poultice was applied to the wound; quinine and opium in full doses, with ale, were given thrice daily.

April 20th. Wound greatly enlarged, edges ragged, under which pus of a most offensive character burrows, a deep slough showing itself over the whole surface of the sore, pulse 110, cheeks red, tongue dry, skin harsh to the feeling, though the night sweats continue. Bromine was now applied, ℨij. of bromine, ℨj. of alcohol, and water q. s. to make four ounces of the mixture.

April 21st. The wound has grown to be three by two inches in size, filled with ash colored sloughs, and the odor intolerable; pulse 110, cheeks red, tongue dry and white coat in center; stools reduced in regard to frequency, yet of a dark color and offensive odor. The same application as the day before.

April 22d. The sloughing enlarged the wound very much, leaving, stretched across its center, a portion of the skin blackened and dead; this, with the edges of the wound, was trimmed off with scissors. The slough covering the center of the sore was deep, and whitish in color. Patient reduced very much; pyæmic symptoms of a most aggravating character, pulse 105, tongue dry and red, stools same as the day before.

Citr. ferri et quiniæ, gr. v., after each meal, or three times a day, was ordered. The bromine was applied, ℨj. to ℨj. of alcohol.

April 23d. The surface of the wound extends three by four inches in size, under the ragged edges of which the fetid pus burrows to the extent of an inch and a half. The surface of the wound still covered with ragged, dark slough, acrid and offensive. Bromine, nearly full strength, was now used over the whole surface of the wound, and injected well beneath its edges, after which the wound was dressed with simple cerate. No change to be noticed in the patient's physical condition.

April 24th. Discharge not so great from the wound, the fetor disappeared, and the sloughs begin to detach themselves. Patient feels better.

April 25th. One-half of the sloughs have come away; pulse 95, tongue more clean, appetite good, slept well last night, the night sweats ceased.

May 3d. Sloughs all gone, the discharges bearing a natural character. The granulations look red and lively. Up to this time the bromine has been used continuously, but was this day discontinued, and a dressing of simple cerate was thought sufficient; pulse 90, tongue clean, bowels nearly natural, appetite good, sleeps well. Though he is weak, the patient expresses himself as feeling very well. From this last date the wound rapidly assumed the most healthy character, and nature, in her most rapid and beautiful manner, commenced repairing the parts, while, at the same time, the constitutional condition of the patient was rapidly improving. The first thing to be remarked in this case was that the constitutional symptoms of pyæmia developed themselves before the local symptoms of gangrene. The second: the remarkably rapid manner which the destructive process in the wound went on. Third: that when a comparatively weak solution was applied, no good accrued, but that when the bromine was applied nearly full strength, not only did the sloughing immediately cease, as also the fetid odor, but the dead portions began to detach themselves, and, moreover, within thirty-six hours, the pyæmic symptoms began to cease. To Dr. Strew, the surgeon in charge, the reporter is indebted for many valuable and timely suggestions.

H. R. VAN NOOKE, A. A. S., U. S. A.,
General Hospital No. 11, *Louisville, Ky.*

Joseph W. Richardson, a private of co. D., 3d Kentucky Infantry, was transferred from Hospital No. 5, Nashville, Tennessee, to this hospital, March 12th, 1863. At the time of his admission he was suffering from pain in the left temple, impaired sight of the left eye, constipation of the bowels, and a feeling of general prostration—the effect, as he supposed, of an attack of erysipelas of the face and neck which he had about the latter part of December, whilst on duty as nurse at Nashville. He was treated by Dr. Fischer, under whose charge he was placed, and by the 1st of June was so far improved as to be able to do guard duty at the hospital. He complained occasionally, however, of vertigo, feeling, as he said, as if he was half drunk. July 5th, about 5 o'clock, P. M., while walking in the yard, he fell down in something like an epileptic fit, and was carried into the ward, where he remained for several days in a state of unconsciousness. On the fourth day after the attack, when consciousness had fully returned, he complained of pain and tenderness at a point about three inches below and to the right of the articulation of the lumbar vertebræ with the sacrum. Upon examination, a spot was found about two and a half inches in diameter, of a dark red color, appearing much like an ordinary bed sore. On the next day the spot appeared somewhat larger, of a darker color, with fluctuation and crepitation upon pressure. Dressed with flax-seed poultice.

6th day. Fluctuation and crepitation increased, surface black and puffed up; evidently in a state of mortification. Dr. Fischer laid open the dead mass with a scalpel, when a quantity of fetid gas, with near a pint of dark offensive watery fluid escaped. The odor was like that emanating from mortifying parts. He then injected the cavity with a solution of bromine, one part of the comp. sol. to four of water.

7th day. Removed the dead mass and filled the cavity with lint, saturated with the compound solution.

8th. Removed the remaining mass, and, having obtained some pure bromine, applied it by means of a mop twice a day. When the whole of the dead matter was removed, it left an enormous cavity, little if any less than a pint bowl in size, the whole depth of the glutei muscles having sloughed away, leaving three or four square inches of the ilium and sacrum exposed.

9th. Sloughing seemed arrested over the greater portion of the surface, but at a few points appeared to be still going on. Bromine continued.

10th. Sloughing arrested, and granulation appearing; sol. brom. comp. ℨj., water ℨ. iv., substituted for pure bromine.

11th. Disease seemed entirely arrested, healthy granulations appearing abundantly. Weak solution of the bromine continued.

The general treatment was supporting. From this time forward the case has improved as rapidly as could be desired, the cavity filling up finely.

Was this a case of hospital gangrene, or of ordinary mortification? At first it presented all the features of ordinary gangrene, but after the mortified mass had been removed, the surface had the appearance and odor of hospital gangrene. What could have caused it? There had been no hospital gangrene in the house for two months, and never any in the ward in which this patient was. The parts might have been bruised while the patient was in a fit. At any rate, the diseased action seemed to yield promptly to the application of the pure bromine, so soon as the dead mass was removed.

J. A. DOUGHERTY, A. A. S., U. S. A.,
General Hospital No. 3, Louisville, Ky., Aug. 6th, 1863.

Greenberry Clayton, aged 24 years, a private of company D, 84th regiment Illinois volunteers, was wounded on the 31st of December, 1863, at the battle of Murfreesboro, Tennessee, by a musket ball striking the outer part of the middle of the right thigh, ranging downwards and inwards posterior to the femur and principal vessels, and passing out through the adductor muscles, making a very extensive and ragged flesh wound. He was admitted to Hospital No. 8, Louisville,

Kentucky, on the 25th of February, 1863, with the wound very painful, parts much swollen, and general health much impaired. The wound at the point of exit of the ball was filled with a dark, tenacious, thick slough, emitting an intolerable fetor, and discharging a thin, sanious matter; a dossil of lint, saturated with pure bromine, was applied directly to the wound, the slough having been first carefully removed with the scissors and forceps. The application was very painful for a few minutes, but the deep seated pain previously existing in the parts was much relieved, and has continued to subside ever since, and at present is all gone. On the next day another application of bromine and water, equal parts, with two parts of bromide of potassium, was applied as before, and suffered to remain on the part fifteen minutes, followed by a linseed poultice. On the third day the gangrene was arrested, the putrid smell disappeared, the swelling rapidly subsiding, and the wound secreting white pus. A dressing of ung. resinæ was applied, and the wound is now rapidly filling up with healthy granulations. Has had no other treatment beyond that of attention to the secretions generally. Small doses of quinine, light but nourishing diet, with a little ale two or three times daily. In this case, although the spread of the gangrene was materially checked by the first application of the bromine, yet it was not until the third application, its spread was entirely arrested. The secreting surface was washed lightly once a day, for four or five days after the gangrene ceased, with a weak solution of bromine, one part of bromine to sixteen of water, and the ung. resinæ with an emollient poultice to the parts in the meantime. The general health of the patient improved rapidly from the first application of the bromine, and the relief experienced from the aching pain in the limb was gratifying in the extreme. The wound is granulating finely, and bids fair to recover soon, without any permanent deformity or contraction of the limb.

FRANKLIN IRISH,
Surgeon in charge Hospital No. 8, Surgeon 77th Reg. Pa. Vols.
March 12th, 1863.

John Bleckenderfer, a private in company C, 79th Pennsylvania Infantry volunteers, was admitted October 1st, 1863. He had been wounded at the battle of Perryville, and was brought from that place about ten days after receiving the wound. The ball had entered about three inches below the knee on the outer part of the leg, and had made its exit about the same distance below the knee on the inside of the leg, making a wound of about six inches in length. The tissues involved were muscular, cellular, and the bones to a slight degree, as the ball had in its course scaled off a small portion of the fibula. The portion of bone involved was very small, and but a small piece of it was extracted, which was done after about one month, at which time it became thoroughly detached. Cold water dressing was applied to the wound, and under its influence the wound healed very rapidly.

On or about the 1st of March the wound, which was then entirely healed over, commenced to open again, the sloughing extending along the track of the old wound and burrowing among the sheaths of the muscles. The discharge was very copious, thin, containing lumps of cellular tissue, and most horribly fetid. The parts about the wound were very painful to the touch, but not "inflamed." I applied at first a mixture of creosote as an injection, hoping that its known antiseptic qualities might produce a favorable change. I did not, however, perceive any advantages from its use; the sloughing still continued, and the odor of the secretion as offensive as before. I then resorted to a solution of bromine, one part of the sol. bromine comp. and one part of water. This gave him a good deal of pain after being applied, but it caused a diminution of pain after the lapse of an hour, the leg feeling much easier than before the application. The effect which it had in destroying the fetid smell was remarkable. That had almost ceased after the second application of the bromine. By continuing this injection the sloughs were rapidly separated, and the secretions became more and more healthy in appearance, the gangrenous smell entirely destroyed, and by the assistance of a soaked bandage and compress the sinuses are rapidly healing. I date the improvement from the first application of the bromine. The result of its use in this, as in every case in which I have used it, is most gratifying.

I am, very respectfully,
Your obedient servant,
A. H. SPEER,
Surgeon 7th Pa. Cavalry, in charge of Hospital No. 9.

To M. GOLDSMITH, *Surgeon U. S. Vols.*

HOSPITAL No 7, MURFREESBORO, TENN.,
April 21st, 1862.

Sir: When you did me the honor to put me in charge of this hospital, you expressed the wish that I would report to you the result of the treatment of hospital gangrene by bromine. I beg leave, respectfully, to submit the report of all the prominent and well marked cases since the 7th March, when I came in charge, and to state that no other treatment has been resorted to in any case.

All of these cases have been seen in their progress and treatment, either by Dr. Brinton, Surgeon U. S. Vols., of Washington, Prof. F. H. Hamilton, Medical Inspector, or Prof. A. C. Post, of New York, or Gunn, of Michigan University. Other cases than these now reported have been treated, but they were slight, or the gangrene not well marked. It is with great pleasure, Sir, that I am able, on the eve of leaving the hospital for the field, to report that not a death has occurred in the gangrene ward, and no case has been treated unsuccessfully.

I have the honor to be,
Sir, your obedient servant,
BENJ'N WOODWARD,
Surgeon 22d Illinois Volunteers.

To G. PERIN, *Surgeon U. S. A.,*
Medical Director Department of the Cumberland.

Report of cases of Hospital Gangrene, treated with bromine, in General Hospital No. 7, Murfreesboro, Tenn., from March 7th to April 27th, 1863, by B. Woodward, Surgeon in charge:

Condor, John, company K, 40th Illinois, was in the hospital March 7th; right leg had been amputated, and gangrene supervened. The stump was open, and the flaps had nearly all sloughed away, leaving exposed the tibia, which exfoliated. The sloughing was arrested by bromine, and the stump dressed as well as it could be. The stump is now nearly healed, only a small spot remaining not cicatrized.

Hall, H. B., company C, 13th Michigan, in hospital March 7th: gangrene from wound of leg; slough was five and a half inches long, two and a half wide, and down to the tibia, the periosteum of which was destroyed, and exfoliation took place. This case is now nearly well.

Bowman, James, company G, 1st East Tennessee regiment, in hospital March 7th: gun shot through left leg, tibia slightly injured, and gangrene filled the wound, which on the outside of the leg was four inches by three, inside five inches by two and a half, and extending through the track of the ball; now nearly well, only a small place on the inside of leg not yet cicatrized.

McCreary, Jesse, 1st East Tennessee, corporal, company C, in ward March 7th: had gangrene of knee from gun shot; had March 7th a large bed sore over the sacrum four inches in diameter, and to the bone, which had sloughed by gangrene and was spreading; now nearly healed up.

Bader, John, company G, 32d Indiana, sent from Hospital No. 1, March 8th, with extensive gangrene of right thigh, from gun shot wound, ball passing through upper third of thigh. On the outside, the wound was five inches in diameter, and destroyed the tensor vaginæ femoris and part of vastus externus; on the inside, the wound was three inches in diameter, and over the femoral artery, which was exposed two and a half inches below Poupart's ligament. The whole of both wounds are now full level with the surface, with healthy granulations, and nearly healed over.

Boyer, William, company B, 49th Ohio, sent from Hospital No. 1, March 12th, with extensive gangrene of left thigh, from a ball passing through. The slough on the outside of the thigh was six inches long by four inches wide, and in some parts one and a half inches deep. The semi-tendinosus and great part of the vastus externus had sloughed away, and a sinus ran down on the outside of the knee. Every part was subjected to the action of bromine, and the sinus injected in its whole length. When he was admitted his death was considered certain, but the progress of the disease was arrested, and now the wound is only three and a half inches long by one inch wide, and filled nearly to the surface with healthy granulations.

Miller, John B., company B, 101st Indiana, sent from Hospital No. 1, April 1st, with gangrene of thigh from gun shot. This case was seen by Professor F. H. Hamilton, when first brought in and before it had been dressed. So extensive was the gangrene that little hopes of his recovery remained. The sore is now only about three inches over, and rapidly filling up with healthy granulations.

Webber, Henry, company F, 7th Pennsylvania Cavalry, sent from Hospital No. 1, April 1st, with gangrene of the foot, from gun shot through part of foot; one toe had been amputated. This case was also seen by Professor Hamilton; now nearly healed.

Biggle, Michael, company G, 101st Indiana, sent from Hospital No. 1, April 12th, with gangrene of the stump after amputation of leg. Flaps had nearly all sloughed away, and the tibia bared. This case is now rapidly healing from granulation, but his health is very bad from remittent fever, and his recovery is doubtful.

Clark, Robert, company G, 101st Indiana, sent from Hospital No. 1, April 12th: gun shot wound through left leg, tibia badly shattered, both wounds of entrance and exit, and track of ball, a mass of gangrene. Both this case and the one following were watched by Professors Post, of New York, and Gunn, of Michigan University, for two days, and the action of the bromine marked. The destruction of parts was promptly arrested, and now the whole wound is rapidly filling up with healthy granulations.

Wilson, L. D., lieutenant, company G, 101st Indiana, sent from Hospital No. 1, April 12th, with extensive gangrene of thigh from gun shot wound. The slough was very extensive and deep, and the whole was in very bad condition. It is now rapidly healing and his general health improving.

In all cases of extensive gangrene, there is remarkable depression; and not the least gratifying feature of the bromine treatment, especially where its *constitutional*, as well as local use is carried out, is the rapid general improvement of the health. The countenance loses its anxious expression, and the leaden, sallow hue is replaced by a healthy, vigorous color, and the appetite is always improved.

Very respectfully,
B. WOODWARD,
Surgeon in charge.

LOUISVILLE, KY., May 18th, 1863.

Philip Dill, private, company D, 13th Louisiana Infantry, aged 35 years, and of good constitution, was wounded at the battle of Stone River, on the 2d of January, 1863, by a rifle or musket ball, which passed through the left thigh, fracturing the femur near the junction of the upper and middle third. There was little laceration of soft parts, and the vessels were uninjured; the leg was amputated January 3d, 1863, at hospital in rear of Murfreesboro, by Surgeon Rief, of 21st Wisconsin, at a point immediately above the injury. The soldier did very well afterwards, and had a good stump, which was entirely healed on March 26th, 1863, at which time he was transferred to this city. The wound assumed a gangrenous character, to some extent, about the 18th of April.

On the 20th the cicatrix had given away, the flaps gangrenous, looking dark and flabby. The fetor was very great; pulse 120 and feeble; tongue dry; cadaverous expression of the countenance, dusky hue of the face, no appetite, singultus, profuse colliquative sweats, and extreme muscular prostration.

April 22d. The sloughs were cleanly removed with a knife, the parts dried, and pure bromine applied to the surface as directed by Surgeon M. Goldsmith, and stimulants, tonics, and nutritious diet given internally.

April 23d. Pulse 100 and stronger; sweats disappeared; no singultus; patient expressed himself as feeling much better; some fetor present in the wound, and two points were retouched with bromine, after which all fetor disappeared, and has not returned up to this date, May 18th, 1863. Granulations began to make their appearance April 18th, four days after the bromine was applied, when the gangrene was arrested and the sloughs entirely separated. The bone protruded about one inch, and is now necrosed.

For several days, about this time, the patient improved rather slowly, the granulations seeming weak and flabby. I attributed this condition to bad ventilation, and advised that he be removed to Hospital No. 1, where he could have better air, which was done, and he improved much more rapidly.

He was much emaciated, and had decubitus; a large bed sore was found over the sacrum, about three inches in diameter, which became gangrenous May 2d. The dead mass was removed, parts dried, and pure bromine applied to the surface. The first application arrested the process, and the sore is now nearly filled up with healthy granulations.

A fact will be noted here that this wound became gangrenous at the same time that the stump was granulating, and not affected in the least by the upper sore. It will also be noticed, as was observed and commented upon by Drs. Barnum and Gregg, who had the care of the case, how rapidly the constitutional symptoms gave way to the effect of the local remedy. After he was removed to Hospital No. 1, he was put in the upper ward, and near a large window, when his improvement seemed to receive a new impetus; since then he has recovered very rapidly, and at this time, May 18th, he has a good appetite, pulse 85, tongue clean and moist, countenance bright, wounds full of granulations, having the appearance of healthy sores.

G. R. WEEKS, *Surgeon U. S. Vols.*
To M. GOLDSMITH, *Surgeon U. S. Vols.*

LOUISVILLE, KY., May 18th, 1863.

Jefferson Terrell, private company K, 17th Tennessee Infantry, C. S. A., aged 20, of good constitution and temperate habits, was wounded in battle of Stone River, December 31st, 1862, in the calf of right leg, carrying away both the tibia and fibula; leg was amputated at the junction of the upper and middle third, at hospital in Murfreesboro, by a Confederate surgeon, January 1st, 1863. He had a good recovery, and he states that his stump was entirely healed. He received a fall, March 13th, by the breaking of his crutch, falling on the end of his still tender stump. On this day he was removed from Murfreesboro to Hospital No. 2, in this city. According to his statement, the stump became gangrenous the next day after his arrival, March 13th. He was treated from that time until 16th April by Surgeon Ronald, 34th Ky. Volunteer Infantry, in charge Hospital No. 2. Surgeon Ronald being relieved at this time, has since been absent, and, as no record of the case was kept by him, it is impossible to know how the case was treated.

On the 16th of April the soldier was attacked by profuse hemorrhage from the stump, the anterior tibial artery having sloughed. At this date the patient was seen by Surgeon Goldsmith, U. S. Vols., who directed me to tie the artery on the face of the slough, and to take charge of the case. I found him cold and pulseless from loss of blood; he had bled about fifty ounces, but the bleeding had been arrested previous to my arrival by the application of a tourniquet to the femoral artery. I ordered him to have stimulants, and waited a short time until reaction should take place. Upon the removal of several adhesive plasters that crossed the face of the stump, I found a large gangrenous mass, the parts looking very dark, and emitting a very offensive odor; I removed, with the scalpel and scissors, a double hand-full of putrid sloughs, partly fluid, partly pulpous, partly solid, and in all stages of putrefactive decomposition. After having done this, the tibia and fibula projected two inches, but were not denuded of periosteum, nor were they necrotic. I prepared the wound as described in accompanying report on hospital gangrene, and applied pure bromine; I also injected the latter into the spaces and cavities, using the precaution to mix it with all the pulp or pultaceous fluid. I then loosened the tourniquet, and no hemorrhage re-appearing, I determined to follow Guthrie—"never to tie an artery unless it bleeds." I gave the necessary directions to avoid loss of blood in case the hemorrhage should recur, and started back to the office, but was soon overtaken by a messenger who informed me that the man was bleeding again, and that the surgeon in charge wished me to return. He had lost one or two ounces only. I then tied the anterior tibial artery as directed, and again applied bromine to the entire surface of the wound, also to the portion of the artery included within the ligature. My reason for so doing was, that I thought the blood might have washed the bromine first applied from the surface, a reason which I now think was insufficient. I ordered a trusty attendant to be kept at the bedside day and night, instructed in the use of the tourniquet, which was loosely applied.

April 17th. Pulse 100, heat of surface about natural; tongue moist; no saccharine odor of the breath; appetite returning; no fetor in the wound; leaden hue of the face disappeared, rested well during the night. Ordered egg-nog, iron, also nutritious diet, and that the wound be kept from the atmosphere.

April 18th. Pulse 100, strong, and full; symptoms everyway better; treatment continued.

April 19th. Pulse 95; warm water dressings to the wound.

April 20th. Still improving, points of granulation springing up; looks pale; no fetor; applied bromine one part, water twelve parts, twice a day. Stimulants continued, decreased in quantity. This evening ligature came away; no hemorrhage from the wound; granulations more abundant but still feeble.

April 21st. Pulse 90, full and soft; patient has rallied very much; pus discharged is inodorous, but is thin and abounds in molecular matter rather than pus globules.

April 22d. Had been removed to Hospital No. 1, which is better ventilated, from which he received much benefit, and from that time he has steadily and rapidly improved up to the present.

May 18th. The tibia and fibula are both covered with healthy granulations, the wound is filling up, and his health is sufficient to enable him to sit up the greater part of the day. He is now fully convalescent.

G. R. WEEKS, *Surgeon U. S. Vols.*
To M. GOLDSMITH, *Surgeon U. S. Vols.*

UNITED STATES GENERAL HOSPITAL, No. 1,
Louisville, Kentucky,
A. C. T. WORTHINGTON, *A. A. Surgeon in charge:*

William Turner, private, company E, 8th Kentucky Volunteers, (a discharged soldier,) was wounded at the Soldier's Home, by an accidental discharge of a pistol in the hands of a comrade, on the 7th of March. The ball entered the anterior and middle portion of the right leg, passing down the tibia about two inches, where it was found and extracted on the 10th of March. Three days after he was admitted into this hospital, he was attacked with inflammation of the right lung (his constitution being much shattered by previous disease), which raged for about ten days. Notwithstanding the supporting treatment, which was strictly observed, he sank into a hectic condition. The harrassing cough, drenching sweats, and the sloughing of the wound, rendered his case almost hopeless. Quinine, iron, whisky, ale, boiled eggs, and beef tea, were freely administered, and stimulating applications were made to the part without effect. On the 1st of April the sloughing had extended to about four inches in length and three in breadth. The soft parts were converted into a greenish grey, pulpy, tenacious mass; the tibia was denuded; there was considerable burning and lancinating pain in the parts, with considerable fetor. The same treatment was continued, with the same result, up to the 11th of April, at which time the sloughs were cut away, the wound thoroughly cleansed, and the solution of bromine was injected. Lint, saturated with the solution, was placed in the wound. The limb was enveloped in a piece of lint, wrung out of tepid water, and covered with oiled silk.

On the morning of the 12th the dressing was removed. The character of the wound was changed from a greyish or blackish stinking mass to a red, healthy looking ulcer, with little or no fetor. A weak solution was again injected, and the lint applied as before.

On the 13th the ulcer still improving; bromine applied to a few parts.

14th. Granulations springing up all over the surface of the ulcer; general condition better; appetite much improved; warm water dressings.

15th. Two small pieces of bone removed; patient entirely free from pain and cough; continue dressings.

17th. Still improving, able to sit up in bed; appetite remarkably good.

20th. Steadily improving since the first application of the bromine, which arrested the gangrene immediately. The patient reports himself in better health than at any time since his discharge. In this case the other usual remedies had a sufficient trial before bromine was used.

[The following case has more relation to the production of pyæmia than to the treatment of hospital gangrene.]

Frederick Jones, private, co. II. 5th. Ky. Infantry, aet. 22.

In the battle of Stone River, Dec. 31st, 1862, he received a gunshot wound in the left leg—the ball entered on the external side of the leg, about six inches below the patella, and, passing inwards and backwards, made its exit on the internal and posterior surface of the leg. According to the patient's statement, the case progressed favorably, until, whilst on his way to Louisville, erysipelas set in.

The disease yielded to treatment, and, as far as the erysipelas was concerned, the patient was well at the time of his admission into this hospital.

Present Condition.—Admitted on the 18th of January, 1863. I find his condition to be as follows: He is pale and somewhat reduced in strength; his intellect is clear, but he is very nervous, and the least excitement, or even muscular exertion, brings on a violent fit of trembling; pulse 104. The left extremity is swollen and red, and the redness and swelling extend above the knee; the wound looks inflamed, the edges are tumid and everted, and almost perfectly dry. Both openings are small, and, on inserting the probe, I find the fibula has been fractured. I could feel a small piece of bone, but as the patient was very irritable, I determined to abstain from interfering until the inflammation should subside.

Treatment.—Cold water dressings were applied to the limb, and the following prescription exhibited: Spts. eth., comp. ext. valerian, aa. ʒj., to be taken every three hours, and the patient was kept on low diet. This treatment was continued until January 25th, when the inflammation was, to a great extent, subdued. The cold water dressings were still continued, and he was allowed an ounce of Sherry wine three times a a day, though still kept on low diet. The wound had in the meantime commenced to suppurate freely, and the pus appeared to be of good quality.

January 27. Patient continues to improve; he sleeps well, but his appetite is still poor. He was ordered to take wine, ʒj., with sol. of quin., ʒij., three times a day. He was also put on half diet.

February 1st. Patient improving, but the wound shows no disposition to heal. Midway between the patella and malleolus, and immediately external to the crest of the tibia, a small abscess is to be seen. On laying it open, it is found to communicate with the wound, which secreted thinner and less healthy looking matter; the odor is becoming offensive; the cold water applications are still continued, as the temperature of the limb immediately rises on their being discontinued, which also causes the parts to become red and painful. Prescription: Sol. quin. ʒij., spts. frumenti ʒss., three times a day, with half diet.

February 8th. During the last week he has continued to take whisky and quinine in doses as above, with the addition of tinct. ferri chloridi, gtt. xx., but there has been no marked improvement. During the night he had a violent chill, and when seen this morning his pulse was 100, feeble and irregular; he has lost all desire for food, and on the left foot is a fluctuating swelling, situated immediately over the external cuneiform bone. I laid it open, and found that a sinus extended up along the leg, behind the external malleolus. Another swelling was found immediately behind the internal malleolus, and extending upward in the track of the flexor longus pollicis; this was also laid open, and both abscesses discharged thin fetid matter. The foot had also commenced to swell. The original wound had commenced discharging matter like that in the abscesses, and on probing it to find fragments of bone, I discovered that there was a large cavity in the thickness of the gastrocnemius muscle. I dilated the opening, and found the walls lined with a thick slough; I could touch fragments of bone with a probe, but never succeeded in laying hold of one with the forceps. Quinine, iron, and alcoholic stimulants were freely administered in doses as stated, and creasote ʒj., alcohol ʒss., water ʒj. ss., sprinkled on poultices applied to the parts; anodynes were given every night, and without them he could not sleep. The following prescription was used: Sol. morph. ʒj. and spts. eth. comp. ʒij. every night at bed-time.

Feb. 15th. The patient is decidedly worse than last week; he has repeated chills, copious night sweats, a loathing of food, especially meat of any kind. Creasote seems to neutralize the fetor, but the sloughs, especially on the foot, are as thick and adhere as firmly as before. During the first two days, after the application of creasote, I think there was some improvement in the appearance of the gangrenous surface, and observed a corresponding improvement in the constitutional symptoms; but he is now in as bad condition as ever. Same treatment continued as before, with the addition of elixir of vitriol in doses of xx. gtt., four times a day.

Feb. 20th. Patient continues in same condition; his pulse is 112 and irregular; chills and night sweats as before. He takes wine ʒj. three times a day, in addition to the prescriptions before mentioned, and oysters and beef tea. The fetor of the discharges is almost intolerable, and the stools emit a similar odor, and are very thin and frequent. I enlarged the original wound, and laid the sinuses on the anterior surface of the leg and on the foot freely open, and injected a strong solution of bromine. This gave the patient severe pain, but changed the character of the discharges almost immediately.

The effects of the application of bromine may be more clearly demonstrated by passing over a few days, and contrasting his condition then to what it was previous to the use of this remedy.

Feb. 28th. He says he feels better; he slept well during the night; has had no chill for several days; night sweats are not so copious, nor do they occur every night. His pulse is 98, and regular; his appetite is returning, but he still has a dislike of meats.

Treatment continued as before, and a weaker solution of bromine injected every twenty-four hours. The fetor has not returned; sloughs are becoming loose, and the discharges have become thicker, and of a healthy color. The opening of entrance is healing rapidly, and this morning I extracted two small pieces of bone through the hole of exit.

March 1st. Patient continues to improve. Wherever the

sloughs have come off, I can discern healthy granulations. On probing the wound, I felt a piece of bone which was extracted, and measured one inch and a half by one half inch.

March 3th. Patient still improving, but very weak. Sores have assumed a healthy appearance. Quinine, iron, and ale, with a generous diet, have been prescribed. Bromine discontinued.

March 6th. Patient feels worse; fetor has returned; discharges have become thinner and of a darker color.

March 7th. He continues getting worse. During yesterday afternoon he was seized with rigors, recurring during the night, and followed by profuse sweats, which have left him quite exhausted. Poultices were applied, and continued until the 13th of March, when the sore on his left foot was again covered with thick sloughs of a dark leaden color, and discharging thin sanious matter, emitting a most intolerable stench. I trimmed off the sloughs with scissors, and injected the compound solution of bromine. This I repeated twice daily, re-applying the poultice after each injection. Tonics and stimulants were freely administered.

The bromine arrested the spreading of the gangrene immediately, and neutralized the stench. On the 18th of March all sloughs had come away, and there was now a healthy granulating ulcer; the swelling, which had been very great, had entirely subsided. I applied unguentum resinæ, and continued to give quinine and alcoholic stimulants. It must be remarked that the constitutional symptoms underwent a marked improvement whenever the gangrenous surface was fully brought under the influence of bromine; in other words, the symptoms of pyæmia disappeared almost simultaneously with the arrest of the fetid discharges, and re-appeared with the return of the latter.

April 2d. The patient is improving rapidly; the ulcer on the foot and leg have entirely healed, but the hole of exit in the calf still remains open. I extracted another piece of bone this morning one inch by one-quarter, and hope, this source of

irritation being removed, the ulcer will speedily follow the example of the other cases. Respectfully submitted,
JOHN A. OCTERLONY, A. A. S. U. S. A.,
General Hospital No. 8, Louisville, Ky.

Hospital No. 4, Nashville, Tenn.,
July 22d, 1863.

Sir: I have the honor to report that, in compliance with your order, I have visited the various hospitals in this city and recorded, on the accompanying blanks, all the cases of hospital gangrene, not before reported, that have occurred therein.

It will be observed that of the thirty-five cases noted, nine proved fatal. This mortality may not be attributed, perhaps in any degree, to the method of treatment employed; for in some of these cases the primary injury was of a serious nature, and the operations performed might have provoked a fatal result, even if gangrene had not existed prior to the operation, or attacked the stump at a subsequent period; but it is a remarkable fact, that in none of the fatal cases was bromine used, except in one, and in this case the bromine was applied only a few hours before the patient's death.

Of the nineteen cases treated with bromine, either from the commencement or during the progress of the disease, all recovered. The remaining seven cases that recovered were treated with nitric acid, antiseptic washes, poultices, etc. The general treatment was similar in all the cases; being of a supporting and tonic character. * * * * *

Average duration of the gangrene in those cases treated with bromine—3 days and 3 hours; average duration of the gangrene in those cases treated without bromine that recovered—5 days and 11 hours.

Very respectfully, your obt. servt.,
C. H. BILL, A. A. Surg. U. S. A.
To A. Henry Thurston, Surg. U. S. V.,
Assist. Med. Dir. Dept. of the Cumberland.

In conclusion, I have to append a consolidated statement of the cases of hospital gangrene reported to me as treated at Louisville, Kentucky; New Albany, Indiana; Nashville, Tennessee, and Murfreesboro, Tennessee. Many more cases have occurred in these several places, but I have been unable to procure such records as would make them suitable bases of tabular statements.

I have carefully excluded all cases of which I am not assured, by personal inspection or the written record, were cases of genuine, unmistakable hospital gangrene. The full records of the cases are reported in the roll sent to the Surgeon General's office, and in the table hereto appended.

It will be seen that among the deaths are four reported as occurring amongst those treated with bromine. In two of these the wounds traversed thick fleshy parts, and, for reasons obvious in the record, the bromine was never applied to the whole gangrenous surfaces or even the major part of them. In another, the gangrenous affection was complicated with acute gangrenous cellulitis (?), extending from the trochanter to the malleoli, the whole cellular membrane underlying the skin having been destroyed. In another the bromine was not applied until the day before death, the man being almost moribund on admission.

Three died of pyæmic affections after the arrest of the gangrene, and the development of granulation throughout the extent of the wounds.

It will be seen, also, that four cases are reported in which the bromine is said to have failed in arresting the disease.

In one of these the bromine had been applied every four hours for several days, and, granulations not appearing, the surgeon applied a weak solution of creasote, and this was followed quickly by the appearance of granulations. In two other cases, after the assiduous application of strong bromine, a solution of the persulphate of iron was used, and the granulations sprang up at once. In another, the surgeon reports that he used the strong applications twice a day for twenty-eight days, and he, too, becoming discouraged, resorted to the use of a cow-dung poultice, and straightway, as in the other cases, the disease was arrested and granulation began. In all of these cases I think it possible that there may have been a too free and frequent use of the bromine. It would be difficult for the cleanest cut surface to granulate with even one application of the compound solution per day, if it was applied in a way calculated to arrest the gangrenous process.

[As these pages are going through the press, a friend suggests that the note on page 32 may convey an impression, neither justified by the facts, nor intended to be made by what is written. The wounded referred to had, all of them, those slight injuries which needed no other dressings than such as were necessary for cleanliness, and such as each man could easily perform for himself. They all, at least such as did not stray away from the command, had their wounds dressed at Bridgeport, the terminus of the ambulance transportation. That their wounds were not afterwards dressed by the medical attendants, was one of those events, unavoidable in the manner of the transportation, on a road given up to troops hurrying to the front; and to which end all other movements were made subservient. I can bear abundant testimony to the fidelity and administrative ability of the accomplished Medical Director of the department, as well as to the industry and efficiency of his staff.]

CONDENSED TABULAR STATEMENT OF CASES OF HOSPITAL GANGRENE.

	Whole Number.	Recovered.	Died.	Amputations.	Average duration of treatment.	Percentage of deaths.
Treated with Bromine, in any way,	152	148	4	0	5 days and 14 hours.	
Treated with Bromine, pure, exclusively,	27	25	2	0	2 do. 23½ "	
Treated with Bromine, in solution, excusively,	86	84	2	0	6 do. 11⅓ "	2.65
Treated with Bromine, pure, after the solution failed,	8	8	0	0	12 do. 18 "	
Treated with Bromine, after Nitric Acid had failed,	23	22	0	1	3 do. 16⅔ "	
Treated with Bromine after other remedies failed,	8	8	0	0	3 do. 4 "	
Treated with Nitric Acid exclusively,	13	5	8	0	3 do. 14⅔ "	61.54 } 50
Treated with other remedies exclusively,	13	7	5	1	7 do. 13⅗ "	38.47
Treated with other remedies after Bromine had failed,	4	4	0	0		

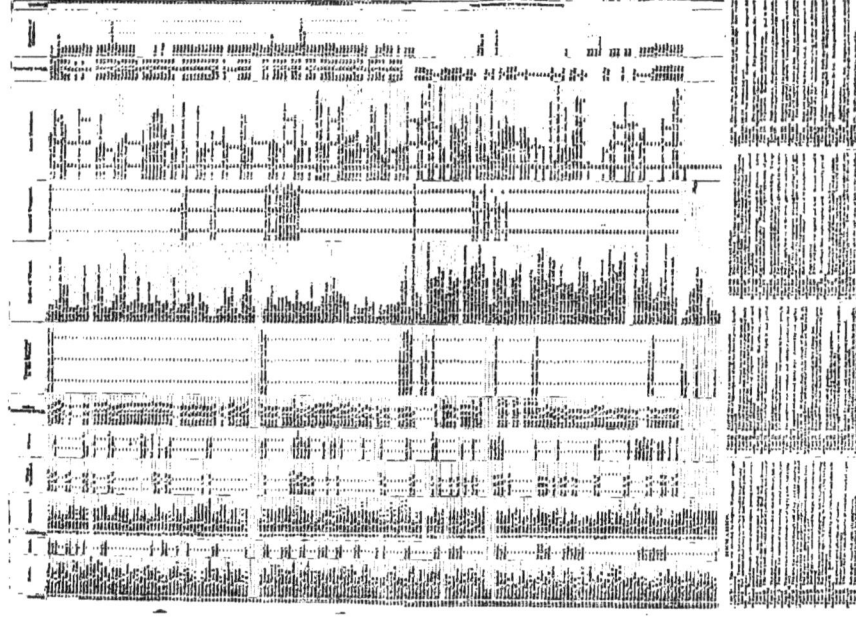

REPORT ON ERYSIPELAS.

The following reports have reference to the influence of the vapor of bromine in arresting the spread of erysipelas in wards, and in the treatment of cases of erysipelas. For the former, the bromine is exposed in the wards in as many places, and in such quantity, as may be found necessary to keep the odor of it constantly present. If the atmosphere is moist, the quantity required will be proportionally large. *This fumigation should be continued as long as the sources of the miasm remain.*

In treating cases of erysipelas with bromine, the following rules are commonly followed:

1st. The face, for example, being the part affected, the parts are to be washed in soap and water, so as to remove all sebaceous matter, and then sponged in clean water.

2d. A mask of patent lint is to be prepared, large enough to cover and extend three inches beyond the erysipelatous area, and in this a ⊥ shaped incision is to be made for the accommodation of the mouth and nose.

3d. Another piece of lint, the exact counterpart of the first.

4th. A piece of oiled silk large enough to cover the face and head, with a like ⊥ incision.

5th. Two pieces of lint large enough to cover the eye-lids are to be smeared with simple cerate and placed over the eyes.

6th. The first mask of lint is to be placed over the face.

7th. The second mask is to be wetted in the following solution:

℞—Bromine, ℨj.
Bromide of potassium, gr. xxx.
Water, ℨx.—M.

—and placed over the first mask.

REPORT ON ERYSIPELAS.

The following reports have reference to the influence of the vapor of bromine in arresting the spread of erysipelas in wards, and in the treatment of cases of erysipelas. For the former, the bromine is exposed in the wards in as many places, and in such quantity, as may be found necessary to keep the odor of it constantly present. If the atmosphere is moist, the quantity required will be proportionally large. *This fumigation should be continued as long as the sources of the miasm remain.*

In treating cases of erysipelas with bromine, the following rules are commonly followed:

1st. The face, for example, being the part affected, the parts are to be washed in soap and water, so as to remove all sebaceous matter, and then sponged in clean water.

2d. A mask of patent lint is to be prepared, large enough to cover and extend three inches beyond the erysipelatous area, and in this a ⊥ shaped incision is to be made for the accommodation of the mouth and nose.

3d. Another piece of lint, the exact counterpart of the first.

4th. A piece of oiled silk large enough to cover the face and head, with a like ⊥ incision.

5th. Two pieces of lint large enough to cover the eye-lids are to be smeared with simple cerate and placed over the eyes.

6th. The first mask of lint is to be placed over the face.

7th. The second mask is to be wetted in the following solution:

℞—Bromine, ʒj.
Bromide of potassium, gr. xxx.
Water, ʒx.—M.

—and placed over the first mask.

8th. The oiled silk is now to be quickly placed over the face and secured by a roller; the object being to cause the vapor of the bromine to come in a dry state in contact with the skin. Some surgeons prefer the use of the solution directly upon the skin.

9th. The application should be renewed once in four hours until the disease subsides.

Sometimes the above formula proves irritating. In that case the strength of the solution must be reduced by the addition of water.

Extract from a report upon the use of bromine in Erysipelas, by B. Woodward, Surgeon 22d Illinois Volunteers, read before the Louisville Society of Army Surgeons:

* * "Though the society has lately honored me by listening to a very imperfect report on bromine, I trust they will pardon me if I again allude to some of the points in that report, while I add such other testimony as I have been able to gain. While erysipelas was in every ward of the Park Barracks Hospital, under my charge, and one ward was full of erysipelas patients, and from which several died, I was ordered by Assistant Medical Director, M. Goldsmith, to procure bromine, and use its vapor in the wards as a prophylactic, and watch its effects. This I did, at first, by dropping bromine into bottles and placing them in the wards, and especially near the erysipelas cases. Although some of these patients were sick from other causes, but one died after the bromine was used.

* * (He died from acute œdema and sloughing of the scrotum.) In most of the cases there was an immediate arrest of the disease, and gradual in all. Not another case originated in the wards while the bromine was used. * * The erysipelas ward, which, for two weeks, had been filled with from twelve to sixteen cases of the worst character, was cleaned, except the walls, which were of unplastered brick, and therefore eminently calculated to absorb the poison; and the room, for forty-eight hours, was kept full of the vapor of bromine, and was then used again as a ward; but not a case has occurred there. * * Bromine, like its congener, iodine, seems to be antagonistic to certain animal poisons; like chlorine, it is an antiseptic; and like both, it is a catalytic(?). If it is objected that this is all hypothetical, we answer it is a new agent. Its status has never been settled, and we have imperfect data from which to reason. The hypothesis, then, is: In bromine we have an agent having controlling power over certain animal poisons, equal to that of iodine, and for certain purposes more available; that it has powers which iodine has not been proven to possess, viz: neutralizing poisons in the atmosphere; and that for this purpose it may be depended upon. Of its constitutional action we know but little, but if its present promise of good holds true, its introduction into the wards of hospitals will disarm them of half their terrors, and to military surgery it will prove an inestimable blessing." *

I can add to Surgeon Woodward's statement the fact, that, at the time when I directed him to procure and use the bromine, the barracks were producing from ten to twenty-five cases of erysipelas per week; and the cases were of terrible severity, more than fifty per cent. proving fatal. No change was made in the number of beds or inmates, no change in the ventilation, and after the lapse of twenty-four hours no more cases occurred, nor did any ever occur, until by some reason the supply of bromine was exhausted, and then a few cases only, but the recurrences were again arrested by the use of the bromine.

During the early part of the spring of 1862, a great many cases of erysipelas occurred at the small-pox hospital then in charge of my brother, A. A. Surg. W. W. Goldsmith. Almost every inmate, nurse as well as patient, was attacked. The stench arising from the cases of small-pox, crowded in low rooms, the utter impossibility of efficient ventilation, engendered their legitimate results. In these circumstances I suggested to him the use of the vapor of bromine as a disinfectant. It was used immediately, and to the effect of at once arresting the spread of the disease. After this, and during the time he continued in charge of the hospital, bromine was used daily in the wards, and no new cases occurred. During the fall and winter, several surgeons in succession having had charge of the hospital, erysipelas again broke out in an epidemic form. The bromine was again used, and with a like result.

During the past winter erysipelas became epidemic in Hospital No. 16, Jeffersonville, Ind. As soon as the bromine was procured and used, the disease disappeared. These are the only occasions occurring under my immediate supervision in which an opportunity was afforded for testing satisfactorily the agency of the bromine in vapor in destroying the miasm producing erysipelas. Beside these, I have had letters from surgeons serving at distant posts, who speak with great confidence of the prophylactic powers of the vapor. The subject needs further investigation. The experiments necessary to test the question are few and simple. All that is it necessary to do is to liberate so much of the vapor of the bromine as is sufficient to make its odor obvious in the infected wards, care being taken that the atmosphere of the wards is as dry as possible.

CASES OF ERYSIPELAS.

The following reports of cases of erysipelas are selected from the records of various hospitals, mostly from those of Hospitals Nos. 19 and 20, in order to show the usual course of events in the cases treated with bromine. Hospitals Nos. 19 and 20 were established for the treatment of cases of erysipelas exclusively. Those occurring at the other hospitals were at once removed to one or the other of these; and almost all the patients were subjected to a uniform course of treatment—that indicated on a previous page.

The rapidity with which the disease was aborted prevented

the development of many grave symptoms, and tended to impress the observer, as perhaps the record will impress the reader, with the idea that they were trivial in their violence. That such impression does not comport with the facts will be seen by the statement, that just at those periods when all the cases in the erysipelas hospitals were apparently so mild, the patients who, by any accident, were delayed for a few days in their removal to these hospitals, presented, on their arrival, all the severities commonly belonging to grave erysipelas of the head and face.

I was called out of hospital to see a gentleman in a private dwelling, who was the subject of erysipelatous inflammation.

This gentleman was born in the State of Vermont; at eight years of age emigrated to Ohio, and grew up there to manhood, then emigrated to Illinois, where he resided before leaving for Nashville, Tenn., to which place he was going when he took sick, and not being able to travel further, he stopped in this city; has had numerous attacks of intermittent fever; is now 56 years of age. I saw him first on the 13th inst.

Face tumefied to the extent of closing both eyes; skin red, with large bullae; inflammation commenced upon the nose, extended upwards and downwards, and spread laterally over the entire face, neck, and scalp; skin between the numerous bullae felt like the grain side of upper leather; bullae filled with dark yellow fluid; severe cephalalgia; says it feels like bursting open; pulse beats 100 per minute, and feeble; bowels costive; urine scanty and high colored; no abnormal appearance elsewhere upon the body; intellect unimpaired; appetite poor; great thirst; no abnormal sounds of heart or lungs.

TREATMENT.

Quinine, gr. ij., every four hours, in ʒj. whisky. Local applications—Bromine 10 drops to ʒj. alcohol—applied as directed by Dr. Woodward.

December 14th. Ten hours after first visit. No discoverable change; appliances moistened again—alcohol ʒj., 10 drops bromine.

Third visit, ten hours afterwards, same day. Still no appreciable alterations, except that the inflammation had ceased to spread. Internal treatment, in addition to quinine and whisky, 12 drops of tincture of opium, and 3 drops tinc. verat. virid., to be given every four hours.

December 15th. Twelve hours intervening since last visit. Patient much better; says he feels like a new man; pulse 60 beats per minute; tongue clammy; no cephalalgia; skin moist; bowels costive; has not had an operation since he left Indianapolis; tumefaction of face, scalp, and neck almost entirely subsided; can now open his eyes; no fluid discoverable in bullæ; redness of skin entirely subsided. Treatment with bromine as before; discontinue laudanum and veratrum viride. One ounce of salts; continue quinine and whisky.

Same day, ten hours since former visit. Bowels have operated freely; patient still improving; pulse now 70. Ordered laudanum and veratrum as before; continue quinine and whisky; apply bromine as before.

December 16th. Fourteen hours since last visit. Pulse 58 beats per minute, full and soft; skin soft and moist; tongue clear, skin covering inflamed surface assuming its normal appearance, except that part occupied by bullæ, the cuticle of which looks white and lies loose upon the true skin. Continued quinine and whisky; continued, also, laudanum and veratrum; discontinued bromine.

December 17th. Patient still improving; says he feels able to start home; continue treatment.

December 18th. Patient fully convalescent, and treatment discontinued, except such covering as is necessary to exclude the air and keep the parts comfortably warm. I dismissed this case to day cured, and he is now on his way home. This was a severe case of erysipelas, and much might be said in favor of the influence exerted by the bromine in the case; but at present I will not speak of its virtues, as I wish to make further tests of its ability to meet and destroy poison.

R. L. STANFORD, *Surgeon U. S. V.*

Case of Erysipelas successfully treated with bromine.

Henry H. Bollis, aged 22, private company F, 18th Ohio, enlisted September 17th, 1861; places of service, Mississippi, Alabama, and Tennessee.

Previous History.—He has been sick for several months previous to admission into Hospital No. 8, suffering from general debility, with great disorder of the digestive organs.

Present Condition.—Admitted into Hospital No. 8 on the 13th of March, 1863. The patient is somewhat emaciated, and very feeble. He looks quite exsanguious, and complained of wandering pains throughout the whole body; he has no appetite; his bowels are irregular, at present constipated. His pulse is 98 to the minute, small and irregular; his tongue is coated with thick white epithelium, but quite moist; a large elevated blotch, decidedly erysipelatous in character, is seen on the right cheek and extending over the right side of the forehead, and also involves the right ear and a portion of the scalp.

Treatment.—Saline purgatives were administered daily, so as to effect, at least, one full passage per diem. Bromine was kept constantly evaporating in the ward, so that its characteristic odor was always distinctly perceptible. The local administration of bromine was resorted to from the beginning; and the following prescription exhibited:

R—Bromine gtt. ix.
Glycerine, ʒj. ss.
M.

Two teaspoonsful to be taken every four hours. This gives one and a-half drops to the dose. Within twenty-four hours (the local application of bromine having been renewed twice during that time) the disease was arrested, and a marked improvement in the constitutional symptoms was also observed. The patient continued to improve steadily, and had entirely recovered on the 16th inst., when the internal administration of bromine was stopped. The air in the ward was still kept heavily impregnated with the vapor of bromine until the 18th inst. This was done to prevent the possibility of the recurrence of the disease, this having taken place in two instances, when the use of the vapor of bromine was discontinued at too early a period.   J. A. OCTERLONY, A. A. S. U. S. A.

[The following cases, reported from Hospital No. 19, were recorded by Mr. Palmer, Medical Cadet.]

William ————, company E, 31st Ohio, age 38, admitted March 26th, 1863, to Hospital No. 19, Ward 2.

Symptoms on Admission.—Been sick seven months, had had chronic diarrhea four months, and during the whole seven months had dyspepsia; great tenderness over abdomen and epigastrium, existing for six months; evidently has chronic inflammation of stomach and intestines; he has vomited a good deal after taking food; appetite poor, but asks for fresh beef and milk; erysipelas extending from median line of nose up to eye-brow, below angle of jaw posteriorly to ear—all on left side; no nervous symptoms, but somewhat emaciated, and very weak; urine normal, bowels quite loose, pain in region of kidneys. Apply bromine to diseased parts. ℞—opii. pulv., grs. ss., at night.

March 27th. Sixteen hours from first application of bromine the erysipelas was arrested; pulse 75; symptoms generally better; continue bromine to face. ℞—Wine, tablespoonful every two hours.

March 28th. Erysipelas subsiding fast; has slight diarrhea; one more application of bromine. ℞—Pul. opii, grs. ss., after every discharge.

March 29th. Erysipelas gone; other symptoms somewhat improved. To drink one bottle porter to day.

March 30th. Convalescent.

GENERAL HOSPITAL, No. 19.,
January 6th, 1863.

Ebenezer McDonald, æt. 33, private, co. I, 87th Indiana Vols., was admitted to Ward No. 4, Hospital 19. On admission he had erysipelas of the face, on left side, extending from median line to occiput, above to frontal suture, and below to angle of jaw; pulse 74, feeble; skin soft, perspiring; tenderness on pressure over abdomen; bowels torpid; icterus well marked; scalding sensation on micturition; rested poorly; rational, and no nervous symptoms; face livid, slight fur on tongue, sordes on teeth and gums, urine high colored.

Treatment.—Bromine, gtt. xx., alcohol ʒj., applied externally, in vapor; bromine, gtt. one-half, internally, every four hours.

Jan. 7th. Face pale, pulse natural, teeth cleaning, abdomen less tender, urine more natural; continue treatment.

Jan. 8th. Erysipelas nearly all gone; skin natural, teeth cleaning, less pain in micturition, urine nearly normal. Stop bromine and give quinine and whisky.

Jan. 9th. Erysipelas gone; icterus has gradually subsided from the beginning of treatment; abdomen free from tenderness; pulse natural; pain in micturition gone. Discharged from service February 25th, 1863.

Daniel Russell, æt. 30, private co. H, 14th Michigan Vols., nurse; attacked March 2d with diphtheria, with slight fever; pulse 75, full; skin moist and warm, tongue slightly coated, bowels regular, urine normal. To-day nitrate of silver, grs. x. to ʒj. of water, was used as a cautery in throat, to take gtt. x. every four hours of tinct. ferri mur., with saturated solution of chlorate of potassa ad libitum. No swelling existed to-day in throat.

March 3d. To-day solution of nitrate of silver grs. xxx. to ʒj. water, used three times; continued iron and chlorate of potassa; swelling beginning to manifest itself; symptoms generally same as yesterday.

March 4th. Swelling continued; same treatment continued. To-day erysipelas has manifested itself on left cheek, extending from medium line of nose, and one-half inch below zygomatic process up to scalp. Bromine, gtts. xl., in saturated solution of chlorate of potassa, was applied in vapor.

March 5th. Erysipelas had not spread any, but is literally killed. The bromine was applied yesterday and during last night only twice; one more application ordered to-day, and other treatment continued.

March 6th. Bromine stopped; other treatment still continued, except the nitrate of silver. This treatment, viz: iron and chlorate of potassa, was continued for several days, and to-day, March 11th, the patient is convalescent.

Milo Butler, co. I, 85th Ill., æt. 29, admitted April 7th, 1863.

Symptoms on Admission. Eight days ago commenced to convalesce from small-pox. Erysipelas began five days ago. It involves the whole right side of the face; aphonia exists; functions normal. Bromine applied in vapor. ℞—Tinct. ferri mur., gtt. x., every two hours.

April 8th. Erysipelas arrested. Continue treatment.

April 9th. Erysipelas gone. Stop treatment above, and give cinchonæ sulph. and whisky every four hours.

April 10th. Sulph. cinchonæ and whisky every four hours.

April 11th. Convalescent.

Case of Erysipelas successfully treated with bromine.

William Lisenby, æt. 37, private, co. E, 80th Illinois Vols., enlisted early in July last. Place of service, Kentucky and Tennessee.

Previous History.—He has enjoyed good health since he entered the army, until, some weeks previous to his admission into Hospital No. 8, he was taken severely ill with pneumonia, which reduced his strength extremely.

Present Condition.—This patient, admitted into General Hospital No. 8 on the 13th day of March, 1863, is found to be very much reduced; quite anæmic, with a low and irregular pulse, his tongue moist, pale, and flabby; he complains of having had no passage for several days; he suffers severe pain in his right ear, which is red and swollen: the meatus auditorous externus is nearly closed, and the patient was supposed to have otitis. Emollient applications were ordered, and seemed to relieve the pain, although the swelling had increased somewhat, when the patient was seen in the evening, March 14th, 1863. On removing the poultice the swelling is found to have extended over the right side of the face, completely closing the eye of that side; the swelling also involved a portion of the scalp and forehead. The poultice was at once removed, and the local application of bromine substituted for it. The patient was constantly exposed to the vapor of bromine with which the air in the ward was continually impregnated; one and a half drops of bromine to ℥ij. of glycerine given every four hours. The progress of the disease was arrested within thirty-six hours. The remedies were used as above stated until the 19th inst., when the patient was pronounced to be entirely recovered from the attack of erysipelas, though still weak and troubled with a cough and pain in his left side, probably the relics of the pneumonia under which he labored previous to his admission; his appetite is good, and he is rapidly progressing towards perfect health.

[The following is inserted to illustrate an occurrence frequently observed in the cases in which iodine had been applied in the early treatment. The iodine, as is well known, causes the dessication of a layer of epithelium, which, while it remains adherent, forms a coating efficient to the protection of the sensitive parts beneath. Such a covering is a complete barrier to the efficacious application of the bromine. In cases of this sort it was always found necessary to detach this hardened epithelium before the bromine could be applied to any useful result.]

E. M. Bradley, private, co. K., 91st Illinois, æt. 20 years, was admitted December 4th to Hospital No. 12, Ward 2, Bed No. 33.

History.—Was born in Massachusetts; lived in Vermont until twelve years of age, since when he has lived in Illinois. He was attacked with diarrhea ten weeks since, followed with chills, which was the fore-runner of bilious fever; was sick at that time eight weeks. He was moved from near Lebanon Junction to Hospital No. 12, in this city, December 4th, with diarrhea, which he had more or less for the whole time; was well previous to enlisting; was sworn into the U. S. service September 8th, 1862.

Present State.—December 11th, 1862. There has been nothing unusual in this case until to-day; his diarrhea has yielded kindly to astringents, &c.

To-day I find his face swollen full (unable to open either eye), red and shining, very painful on pressure; slight diarrhea. Diagnosis: erysipelas.

December 11th. Face, cheeks, and lips are very much swollen; do not pit on pressure, but have the feeling peculiar to erysipelas; skin dry; pulse 95 per minute—beats full and strong; tongue—edges red, light coating through the center; intellect and nervous system normal. The swelling extends to the forehead, nearly to the hair, and is exceedingly tender; abdomen—slight tenderness on pressure; urine scanty and high colored.

Prescription, Diet, &c.—Whisky, half ounce, quinine two grains, every four hours. Paint surface for two inches beyond inflamed surface with tr. iodine. Extra diet.

December 12th.—Cheeks and lips more swollen; can not see with either eye; skin dry; pulse 100; tongue red, coating through center darker than yesterday; restlessness; the swelling has extended into the scalp, pits on pressure, great tenderness; abdomen less tender than yesterday; urine more abundant.

Prescription, Diet, &c.—Continue whisky and quinine; remove hair and whiskers; paint whole head with tr. iodine; five grains Dover's powder at night.

December 13th. Swelling increased, more painful than yesterday; skin dry; pulse 100; tongue, edges darker and redder than yesterday; more restlessness; swelling and tumefaction spreading to whole scalp; slight diarrhea; urine normal.

Prescription, Diet, &c.—Omit tr. iodine; apply bromine; continue whisky and quinine; give tinct. opium, 16 drops, tr. veratrum viride, 4 drops, every six hours, to check diarrhea.

December 14th. Face, cheeks, and lips—swelling extending; skin dry; pulse 85; tongue moist and dark; sleeps better; no change. *The skin is covered with cuticle hardened by tr. iodine;* less diarrhea; urine normal. *Remove cuticle*, which tears off easily. Continue treatment.

December 15th. Face, cheeks, lips—swelling less, can see from left eye; skin moist; pulse 80; tongue moist and lighter than yesterday; intellect normal; less swelling and pain; abdomen and urine normal. Continue treatment.

December 16th. Less tumefaction of face and cheeks; skin moist; pulse 80; tongue cleaning; intellect normal; swelling and tenderness nearly gone; sloughing of right upper side: abdomen and urine normal. Omit bromine, tr. opii, and veratrum viride. Continue whisky and quinine.

December 19th. Face, cheeks, and lips—slight tumefaction; skin moist; pulse 80; tongue clean; intellect normal; abscess in left upper and right under eye-lid; abdomen and urine normal. Continue whisky and quinine; open abscesses; apply poultices.

December 21st. No tumefaction except above eye-lid; skin moist; pulse 80; tongue clean; intellect normal; ulcerations and abcesses about eyes filling up with healthy granulations; abdomen and urine normal. Continue treatment.

December 25th. Face, cheeks, and lips natural; skin moist; pulse 76; tongue clean; intellect normal; ulcerations nearly filled; abdomen and urine normal. Omit all treatment except extra diet.

The consolidated tabular statement of the cases of erysipelas, treated at this place, is annexed. My own observation of the cases convinced me that the earlier cases, those treated before the adoption of the bromine, were much more severe than those occurring at later dates. Still every case recorded was one of well marked and genuine erysipelas. The roll of these cases, and the detailed records, are in the Surgeon General's office.

In conclusion, I beg leave to state that none of the cases of hospital gangrene or of erysipelas, the records of which are given here, or have been heretofore forwarded to the Surgeon General's office, were treated by me. The results here presented are those arrived at in the practice of some sixty medical officers stationed at this and at some distant points. Those serving here were under my immediate supervision. Those at other points had, many of them, but an imperfect acquaintance with the reasoning upon which the treatment was based*; nor were they acquainted, many of them, fully with the clinical measures necessary for the most efficacious application of the bromine. My own views on this point were not sufficiently set forth in the circular issued in the beginning of these investigations; nor had I then definitely determined in my own mind the points which I have striven to enforce in the foregoing pages.

As the measure of the value of a remedy is, in some sense, in its working value, not to be determined so much by its effects in the hands of gifted experts as in the hands of the generality, so to speak, the average of practitioners, I have preferred to present, for the consideration of the Surgeon General, such records as would exemplify the aggregate of the observations thus far made.

I beg, also, to remind the Surgeon General that the results thus far reached are results had with a new remedial agent, the rules for the use of which had to be worked out in the midst of many difficulties, not the least of these being the traditional indisposition to abandon the tried resources of medicine in favor of new agents based upon views not in exact accordance with accredited dogmata.

I have the honor to be, your obedient servant,

M. GOLDSMITH, *Surgeon, U. S. V.*

TABULAR STATEMENT OF CASES OF ERYSIPELAS, TREATED IN GENERAL HOSPITAL AT LOUISVILLE, KY.

	Number of cases.	Recovered.	Died.	\multicolumn{12}{c}{RECOVERIES COMPLICATED WITH.}	\multicolumn{8}{c}{DEATHS COMPLICATED WITH.}	Average date at which the spread of erysipelas was arrested.	Average date at which the erysipelas disappears.	Average date of convalescence.	
				Wound.	Diphtheria.	Suppuration of the cellular tissue.	Pneumonia.	Phthisis.	Typhus.	Œdema Glottidis.	Chronic Diarrhea.	Gangrene of the tongue.	Secondary Syphilis.	Chronic bronchitis.	Uncomplicated.	Typhus.	Phthisis.	Gangrene of the arm.	Typhoid pneumonia.	Œdema Glottidis.	Tetanus.	Secondary hemorrhage.	Uncomplicated.				
No. 1. Treated without bromine,	63	31	32	...	...	...	...	...	...	...	...	...	...	...	31	1	...	...	1	1	...	...	29	8 days 6 hrs.	19 dy's 21 hrs.	19 dy's 21 hrs.	
No. 2. Treated in any way with brom'e	165	146	19	4	2	1	1	1	1	1	6	1	1	3	124	10	1	1	1	...	1	...	5	4 days 12 hrs.	6 days 8 hrs.	8 days 14 hrs.	
No. 3. Treated with bromine topically, & with vapor in ward,	104	99	5	4	2	1	1	1	1	1	6	1	1	3	77	3	1	...	1	...	...	...	...	3 days 1 hour	5 days 8 hrs.	7 days 2 hrs.	This class is included in class No. 2.
No. 4. Since last report treated with bromine topically	58	55	3	...	...	...	...	...	3	...	...	...	...	...	52	2	...	...	...	...	...	1	...	1 day 8 hrs.	3 days 22 hrs.	5 days 5 hrs.	Cases treated since last report, mostly in hospitals Nos. 19 & 20.
Totals............																											

www.ingramcontent.com/pod-product-compliance
Lightning Source LLC
Chambersburg PA
CBHW020900160426
43192CB00007B/1014